Standards, Conformity Assessment, and Accreditation for Engineers

STANDARDS, CONFORMITY ASSESSMENT, AND ACCREDITATION FOR ENGINEERS

ROBERT D. HUNTER

CRC Press
Taylor & Francis Group
Boca Raton London New York

CRC Press is an imprint of the
Taylor & Francis Group, an **informa** business

CRC Press
Taylor & Francis Group
6000 Broken Sound Parkway NW, Suite 300
Boca Raton, FL 33487-2742

© 2009 by Taylor & Francis Group, LLC
CRC Press is an imprint of Taylor & Francis Group, an Informa business

No claim to original U.S. Government works
Printed in the United States of America on acid-free paper
10 9 8 7 6 5 4 3 2 1

International Standard Book Number-13: 978-1-4398-0094-2 (Softcover)

**Visit the Taylor & Francis Web site at
http://www.taylorandfrancis.com**

**and the CRC Press Web site at
http://www.crcpress.com**

To the Memory of My Wife
Betty Miller Hunter

Contents

List of Figures

List of Tables

Preface

Standards are ubiquitous! But conformity assessment and accreditation are seldom talked about in ordinary conversation. These terms have been defined and redefined in a new set of International Standards by the International Organization for Standardization/International Electrotechnical Commission (ISO/IEC). My aim in this work is to couple these subjects more tightly than is usually done. I also aim to produce an overview of these subjects including international, regional, national, provincial (state), and company aspects. I have fired one of Kipling's "six honest serving men," the *how* of these subjects; those aspects are too variable and complex to treat here in a meaningful way. Except for some specialized examples not often discussed in the literature, I have steered clear of much of the interesting history of these subjects since that is well treated elsewhere. I have been forced to omit many interesting aspects of these subjects, but hope that they will be revealed by a perusal of the literature cited.

In addition to engineers and standards professionals, this book will be of interest to those involved in regulatory compliance, government, law, business, and trade.

Acknowledgements

Many people have contributed in various ways but special mention must be made of S. Joe Bhatia and Rosemary Maginniss of the American National Standards Institute (ANSI), the International Organization for Standardization (ISO), the International Electrotechnical Commission (IEC), Joanne Saathoff and Emily Sharp of cPubs, and my publisher Nora Konopka of CRC Press and her editors. The following also made significant contributions: Tim Davis, David Edgerly, Mike Floyd, Otto Friedrich Jr., Barbara Hunter, Ken Krechmer, Gary Kushnier, JoAnne Overman, Ross Quan, Gary Schrempp, Gary Sutorius, and Gregory Tassey.

The Author

Robert D. (Bob) Hunter is a licensed professional engineer in Austin, Texas, and has been involved in standards, conformity assessment, and accreditation as a company employee and as a consultant to several technical and computer companies for more than twenty-five years. He has served on standards committees at both international and national levels and served on the Industrial Advisory Committee (IAC) of Underwriters Laboratories (UL) that developed one of the first national standards (UL 1950) that was based on a major international standard (IEC 60950). Hunter has written and lectured on standards-related subjects in his major fields of interest, including product safety, electromagnetic compatibility, and regulatory compliance.

Hunter is a member of the Institute of Electrical and Electronics Engineers (IEEE) and the Association for Computing Machinery (ACM). His consulting firm is a member of the American National Standards Institute (ANSI) and the American Society for Testing and Materials International (ASTMI).

1

ISO/IEC Guide 2 Standardization and Related Activities
General Vocabulary

Introduction

Some writers about standards snap off salutes to Guide 2 and IS 17000, then proceed to ignore most of the terms and their definitions in their presentations. As indicated in the Preface, my aim is to consistently use Guide 2 and IS 17000 vocabulary throughout. The **foreword** of Guide 2 gives some of its history. It was driven by the need for consistency in standards matters, notably in regard to the General Agreement on Tariffs and Trade (GATT), the predecessor of the World Trade Organization (WTO).

The eighth edition of Guide 2 groups related concepts under generic headings. Clauses 1 to 11 concern standardization, proceeding from more general to more specific aspects.

Guide 2 includes an alphabetical index of defined terms for easy lookup. That index is included in Appendix 1 to facilitate the finding of terms and definitions. Abbreviations and acronyms are collected in Appendix 3 for ease of location.

The individual terms (and definitions) in Guide 2 are used as modules to construct higher-level concepts. As an example from the introduction to Guide 2, the term **safety standard** is defined as a **standard** (3.2) concerned with the definition of **safety** (2.5). This concept of modularity extends to standards themselves as is demonstrated in Chapter 3.

Although the original document defines terms in the official languages of the ISO and IEC (English, French, and Russian), it includes terms (without definitions) in German (de), Spanish (es), Italian (it), Dutch (nl), and Swedish (sv).

The importance of ISO/IEC Guide 2, ISO/IEC 17000, and ISO/IEC 17011 (Accreditation) for the understanding of standards, conformity assessment, and accreditation cannot be overstated.

Guide 2 and IS 17000 are referenced in (hence made a part of) a great many international, regional, and national standards; they are the basic foundations on which the terminology of standards and conformity assessment rests. If people want to expand or extend the terms applying to standards, conformity assessment, and accreditation per se, they are, of course, free to do so, but at the risk of creating confusion and losing clear communication. Communication is, after all, what standards are all about.

Chapters 1 and 2 use a two-column format. The left column includes the Guide 2 and IS 17000 English terms and their definitions. The right columns include some commentary that hopefully illuminates the subject and compares Guide 2 and IS 17000 with some Western (mostly U.S.) national terminology. Terms in Guide 2 and IS 17000 are in boldface in the definitions, but later in this text are indicated as defined terms (where emphasis is needed) by initial capitalization.

Guide 2	Commentary
1. Standardization	
1.1 Standardization	
Activity of establishing, with regard to actual or potential problems, **provisions** for common and repeated use, aimed at the achievement of the optimum degree of order in a given context.	
Note 1: In particular, the activity consists of the processes of *formulating, issuing,* and *implementing* **standards**.	As shown in Note 1, this definition is broader than most other definitions. Words in italics here are for emphasis.
Note 2: Important benefits of standardization are improvement of the suitability of products, processes and services for their intended purposes, prevention of barriers to trade, and facilitation of cooperation.	
1.2 Subject of Standardization	
(topic to be standardized)	
Note 1: The expression "product, process or service" has been adopted throughout this Guide to encompass the subject of standardization in a broad sense, and should be understood equally to cover, for example, any material, component, equipment, system, interface, protocol, procedure, function, method or activity.	That pretty well says it all!
Note 2: **Standardization** may be limited to particular aspects of any subject. For example, in the case of shoes, criteria could be standardized separately for sizes and durability.	

Guide 2	Commentary

1.3   Field of Standardization

Domain of standardization (deprecated) group of related **subjects of standardization.**

Note: Engineering, transport, agriculture and quantities and units, for example, could be regarded as fields of standardization.

Deprecated—expressed disapproval of

Information Technology (IT) might be considered a field.

1.4   State of the Art

Developed stage of technical capability at a given time as regards products, processes and services, based on the relevant consolidated findings of science, technology and experience.

A term probably invented by the U.S. Supreme Court in a patent case.

1.5   Acknowledged Rule of Technology

Technical **provision** acknowledged by a majority of representative experts as reflecting the **state of the art**.

Note: A normative document on a technical subject, if prepared with the cooperation of concerned interests by consultation and consensus procedures, is presumed to constitute an acknowledged rule of technology at the time of its approval.

Normative document is defined in 3.1.

1.6   Level of Standardization

Geographical, political, or economic extent of involvement in **standardization.**

The European Norm (EN) is an example of a regional–political standard.

1.6.1   International Standardization

Standardization in which involvement is open to relevant **bodies** from all countries.

1.6.2   Regional Standardization

Standardization in which involvement is open to relevant **bodies** from countries from only one geographical, political or economic area of the world.

The European Union is an example where regional standardization occurs.

1.6.3   National Standardization

Standardization that takes place at the level of one specific country.

In the United States, this takes place at the American National Standards Institute (ANSI).

Guide 2	Commentary
Note: Within a country or a territorial division of a country, standardization may also take place on a branch or sectoral basis (e.g., ministries) or at local levels, at association or company levels in industry and in individual factories, workshops and offices.	ASTMI (American Society for Testing and Materials International) and IEEE (Institute of Electrical and Electronics Engineers) standards are examples.

1.6.4   Provincial Standardization

Standardization that takes place at the level of a territorial division of a country.	The State of Texas Standards for Roadways is an example.
Note: Within a country or a territorial division of a country, standardization may also take place on a branch or sectoral basis (e.g., ministries), at local levels, at association and company levels in industry, and in individual factories, workshops, and offices.	

1.7   Consensus

General agreement, characterized by the absence of sustained opposition to substantial issues by any important part of the concerned interests and by a process that involves seeking to take into account the views of all parties concerned and to reconcile any conflicting arguments.	Typically a percentage of the bodies voting. In the IEC, it is 75 percent.
Note: Consensus need not imply unanimity.	

2.   Aims of standardization

Note: The general aims of **standardization** follow from the definition in 1.1. Standardization may have one or more specific aims, to make a product, process or service fit for its purpose. Such aims can be, but are not restricted to, variety control, usability, compatibility, **interchangeability**, health, **safety, protection of the environment, product protection**, mutual understanding, economic performance, and trade. They can be overlapping.

2.1   Fitness for Purpose

Ability of a product, process, or service to serve a defined purpose under specific conditions.	This definition has a counterpart in the Uniform Commercial Code (UCC): Warranty of Fitness for a Particular Purpose.

Guide 2	Commentary

2.2 Compatibility

Suitability of products, processes or services for use together under specific conditions to fulfill relevant **requirements** without causing unacceptable interactions.

2.3 Interchangeability

Ability of one product, process, or service to be used in place of another to fulfill the same **requirements.**

Note: The functional aspect of interchangeability is called "functional interchangeability" and the dimensional aspect "dimensional interchangeability."

2.4 Variety Control

Selection of the optimum number of sizes or types of products, processes, or services to meet prevailing needs.

Note: Variety control is usually concerned with variety reduction.

The number of colors available in production cars, for example, is usually limited.

2.5 Safety

Freedom from unacceptable risk of harm.

Note: In **standardization**, the safety of products, processes and services is generally considered with a view to achieving the optimum balance of a number of factors, including nontechnical factors such as human behavior, that will eliminate avoidable risks of harm to persons and goods to an acceptable degree.

2.6 Protection of the Environment

Preservation of the environment from unacceptable damage from the effects and operations of products, processes and services.

2.7 Product Protection

Environmental protection (deprecated). Protection of a product against climatic or other adverse conditions during its use, transport or storage.

"Deprecated" means you shouldn't use that expression.

Guide 2	Commentary

3. Normative Documents

3.1 Normative Document

Document that provides rules, guidelines or characteristics for activities or their results.

The **normative references** in this Guide 2 document make ISO/IEC 17000 a **normative document** for this standard.

Note 1: The term "normative document" is a generic term that covers such documents as **standards, technical specifications, codes of practice** and **regulations**.

Note 2: A "document" is to be understood as any medium with information recorded on or in it.

Note 3: The terms for different kinds of normative documents are defined considering the document and its content as a single entity.

3.2 Standard

Document, established by **consensus** and approved by a recognized **body**, that provides, for common and repeated use, rules, guidelines, or characteristics for activities or their results, aimed at the achievement of the optimum degree of order in a given context.

Note: Standards should be based on the consolidated results of science, technology, and experience, and aimed at the promotion of optimum community benefits.

3.2.1 Publicly Available Standards

Note: By virtue of their status as **standards**, their public availability, and their **amendment** or **revision** as necessary to keep pace with the **state of the art**, **international, regional, national**, and **provincial standards** (3.2.1.1, 3.2.1.2, 3.2.1.3, and 3.2.1.4) are presumed to constitute **acknowledged rules of technology**.

Frequently abbreviated as PAS (alternately, publicly available specification).

3.2.1.1 International Standard

Standard that is adopted by an **international standardizing/standards** organization and made available to the public.

In practice that includes mostly the **ISO, IEC**, and **ITU** standards. The World Trade Organization has offered an alternative definition (see Chapter 6 for details).

Guide 2	Commentary

3.2.1.2　Regional Standard

Standard that is adopted by a **regional standardizing/standards organization** and made available to the public.	EN 60950-1 for the safety of IT equipment is an example.

3.2.1.3　National Standard

Standard that is adopted by a **national standards body** and made available to the public.	An ASTMI standard that is "recognized" by ANSI is an example.

3.2.1.4　Provincial Standard

Standard that is adopted at the level of a territorial division of a country and made available to the public.

3.2.2　Other Standards

Note: **Standards** may also be adopted on other bases, for example, branch standards and company standards. Such standards may have a geographical impact covering several countries.

3.3　Prestandard

Document that is adopted provisionally by a **standardizing body** and made available to the public in order that the necessary experience may be gained from its application on which to base a **standard**.	In Europe, the prestandard for a European Norm (EN) is identified as a "prEN" standard.

3.4　Technical Specification

Document that prescribes technical **requirements** to be fulfilled by a product, process or service.

Note 1: A technical specification should indicate, whenever appropriate, the procedure(s) by means of which it may be determined whether the requirements given are fulfilled.

Note 2: A technical specification may be a **standard**, a part of a standard, or independent of a standard.

Note 3: Note 3 applies to the Russian version only.	The Russian language Note 3.

Guide 2	Commentary

3.5 Code of Practice

Document that recommends practices or procedures for the design, manufacture, installation, maintenance, or utilization of equipment, structures, or products.

Note: A code of practice may be a **standard**, a part of a standard, or independent of a standard.

Usually called a "Code" for brevity.

3.6 Regulation

Document providing binding legislative rules, which is adopted by an **authority**.

3.6.1 Technical Regulation

Regulation that provides technical **requirements** either directly or by referring to or incorporating the content of a **standard, technical specification**, or **code of practice**.

Note: A technical regulation may be supplemented by technical guidance that outlines some means of compliance with the requirements of the regulation, that is, a **deemed-to-satisfy provision**.

The "electrical codes" of states, counties, and cities in the United States, usually based on the NFPA (National Fire Protection Association) National Electrical Code (NEC), are good examples.

4. Bodies Responsible for Standards and Regulations

4.1 Body

(Responsible for standards and regulations) legal or administrative entity that has specific tasks and composition.

Note: Examples of bodies are **organizations, authorities**, companies and foundations.

4.2 Organization

Body that is based on the membership of other bodies or individuals and has an established constitution and its own administration.

4.3 Standardizing Body

Body that has recognized activities in **standardization**.

Guide 2	Commentary
4.3.1 Regional Standardizing Organization	
Standardizing **organization** whose membership is open to the relevant national **body** from each country within one geographical, political or economic area.	CEN (European Committee for Standardization), CENELEC (European Committee for Electrotechnical Standardization), and ETSI (European Telecommunications Standards Institute) are examples.
4.3.2 International Standardizing Organization	
Standardizing **organization** whose membership is open to the relevant national **body** from every country.	The term "organization" is frequently replaced by the term "body."
4.4 Standards Body	
Standardizing body recognized at national, regional or international level, that has as a principal function, by virtue of its statutes, the preparation, approval or adoption of **standards** that are made available to the public.	Sometimes referred to as a Standards Developing Organization (SDO).
Note: A standards body may also have other principal functions.	
4.4.1 National Standards Body	
Standards body recognized at the national level that is eligible to be the national member of the corresponding **international** and **regional standards organizations.**	ANSI is the U.S. NSB. Also frequently called a national standards organization (NSO).
4.4.2 Regional Standards Organization	
Standards organization whose membership is open to the relevant national **body** from each country within one geographical, political or economic area only.	An example is the European Union (EU) European Committee for Standardization (CEN). Also called a regional standards *body*.
4.4.3 International Standards Organization	
Standards organization whose membership is open to the relevant national **body** from every country.	ISO and IEC are examples. (Also called an international standards *body*.)
4.5 Authority	
Body that has legal powers and rights.	A national authority in the United States is the Occupational Safety and Health Administration (OSHA).
Note: An authority can be regional, national or local.	

Guide 2	Commentary

4.5.1   Regulatory Authority

Authority that is responsible for preparing or adopting **regulations.**

4.5.2   Enforcement Authority

Authority that is responsible for enforcing **regulations**.

The Federal Communications Commission (FCC) is an example of *both* a regulatory authority and an enforcement authority.

Note: The enforcement authority may or may not be the **regulatory authority**.

5.   Types of standards

Note: The following terms and definitions are not intended to provide a systematic classification or comprehensive list of possible types of **standards**. They indicate some common types only. These are not mutually exclusive; for instance, a particular **product standard** may also be regarded as a **testing standard** if it provides **test methods** for characteristics of the product in question.

5.1   Basic Standard

Standard that has a wide-ranging coverage or contains general **provisions** for one particular field.

An example is IEC 60664, Insulation coordination for equipment within low-voltage systems.

Note: A basic standard may function as a standard for direct application or as a basis for other standards.

5.2   Terminology Standard

Standard that is concerned with terms, usually accompanied by their definitions, and sometimes by explanatory notes, illustrations, examples, etc.

ISO/IEC Guide 2 and ISO/IEC 17000 are examples.

5.3   Testing Standard

Standard that is concerned with **test methods**, sometimes supplemented with other **provisions** related to **testing**, such as sampling, use of statistical methods, sequence of tests.

American Society of Testing and Materials International (ASTMI) E 659, Standard Test Method for Autoignition Temperature of Liquid Chemicals, for example.

Guide 2	Commentary

5.4 Product Standard

Standard that specifies **requirements** to be fulfilled by a product or a group of products, to establish its **fitness for purpose**.

Note 1: A product standard may include, in addition to the fitness for purpose requirements, directly or by reference, aspects such as terminology, sampling, testing packaging and labeling and, sometimes, processing requirements.

Note 2: A product standard can be either complete or not, according to whether it specifies all or only a part of the necessary requirements. In this respect, one may differentiate between standards such as dimensional, material, and technical delivery standards.

5.5 Process Standard

Standard that specifies **requirements** to be fulfilled by a process, to establish its **fitness for purpose**.

5.6 Service Standard

Standard that specifies **requirements** to be fulfilled by a service, to establish its **fitness for purpose**.

Note: Service standards may be prepared in fields such as laundering, hotel-keeping, transport, car-servicing, telecommunications, insurance, banking, trading.

5.7 Interface Standard

Standard that specifies **requirements** concerned with the **compatibility** of products or systems at their point of interconnection.

Institute of Electrical and Electronics Engineers (IEEE) 488 standards are well known examples.

5.8 Standard on Data to be Provided

Standard that contains a list of characteristics for which values or other data are to be stated for specifying the product, process or service.

Note: Some standards, typically, provide for data to be stated by suppliers, others by purchasers.

Guide 2	Commentary

6.   Harmonization of Standards

Note: Technical regulations can be harmonized like **standards**. Corresponding terms and definitions are obtained by replacing "**standards**" by "**technical regulations**" in definitions 6.1 to 6.9, and "standardizing bodies" by "authorities" in definition 6.1.

6.1   Harmonized Standards

Equivalent standards.

Standards on the same **subject** approved by different **standardizing bodies** that establish **interchangeability** of products, processes, and services, or mutual understanding of test results or information provided according to these standards.

Underwriters Laboratories (UL) Standard 60 950-1, Safety of Information Technology Equipment (ITE) is one example.

Note: Within this definition, harmonized standards might have differences in presentation and even in substance, for example, in explanatory notes, guidance on how to fulfill the **requirements** of the standards, preferences for alternatives and varieties.

6.2   Unified Standards

Harmonized standards that are identical in substance but not in presentation.

6.3   Identical Standards

Harmonized standards that are identical in both substance and presentation.

Note 1: Identification of the standards may be different.

Note 2: If in different languages, the standards are accurate translations.

6.4   Internationally Harmonized Standards

Standards that are harmonized with an **international standard**.

6.5   Regionally Harmonized Standards

Standards that are harmonized with a regional **standard**.

Guide 2	Commentary

6.6   Multilaterally Harmonized Standards

Standards that are harmonized between more than two **standardizing bodies**.

6.7   Bilaterally Harmonized Standards

Standards that are harmonized between two **standardizing bodies**.

7.   Content of Normative Documents

7.1   Provision

Expression in the content of a **normative document** that takes the form of a **statement**, an **instruction**, a **recommendation**, or a **requirement**.

Note: These types of provision are distinguished by the form of wording they employ; for example, instructions are expressed in the imperative mood, recommendations by the use of the auxiliary "should," and requirements by the use of the auxiliary "shall."

7.2   Statement

Provision that conveys information.

7.3   Instruction

Provision that conveys an action to be performed.

7.4   Recommendation

Provision that conveys advice or guidance.

7.5   Requirement

Provision that conveys criteria to be fulfilled.

7.5.1   Exclusive Requirement

Mandatory requirement (deprecated).
Requirement of a **normative document** that must necessarily be fulfilled in order to comply with that document.

Note: The term "mandatory requirement" should be used to mean only a requirement made compulsory by law or **regulation**.

Guide 2	Commentary

7.5.2   Optional Requirement

Requirement of a **normative document** that must be fulfilled in order to comply with a particular option permitted by that document.

Note: An optional requirement may be either (a) one of two or more alternative requirements; or (b) an additional requirement that must be fulfilled only if applicable and that may otherwise be disregarded.

7.6   Deemed-to-Satisfy Provision

Provision that indicates one or more means of compliance with a **requirement** of a **normative document**.

7.7   Descriptive Provision

Provision for **fitness for purpose** that concerns the characteristics of a product, process or service.

Frequently called a "design standard" in the United States.

Note: A descriptive provision usually conveys design, constructional details, etc., with dimensions and material composition.

7.8   Performance Provision

Provision for **fitness for purpose** that concerns the behavior of a product, process or service in or related to use.

Generally preferred to a "design standard."

8.   Structure of Normative Documents

8.1   Body

(Of a normative document) set of **provisions** that comprises the substance of a **normative document**.

Note 1: In the case of a standard, the body comprises general elements relating to its subject and definitions, and main elements conveying provisions.

Note 2: Parts of the body of a normative document may take the form of annexes ("normative annexes") for reasons of convenience, but other (informative) annexes may be additional elements only.

Guide 2	Commentary

8.2 Additional Element

Information that is included in a **normative document** but has no effect on its substance.

Note: In the case of a **standard**, additional elements may include, for example, details of publication, foreword, and notes.

9. Preparation of Normative Documents

9.1 Standards Program

Working schedule of a **standardizing body** that lists its current items of **standardization** work.

9.1.1 Standards Project

Specific work item within a **standards program**.

9.2 Draft Standard

Proposed **standard** that is available generally for comment, voting, or approval.

9.3 Period of Validity

Period of time for which a **normative document** is current that lasts from the date on which it becomes effective ("effective date"), resulting from a decision of the **body** responsible for it, until it is withdrawn or replaced.

Usually five years or longer, in some cases ten years.

9.4 Review

Activity of checking a **normative document** to determine whether it is to be reaffirmed, changed, or withdrawn.

Not to be confused with "review" in Conformity Assessment (*revised*), which is another word for "changed."

9.5 Correction

Removal of printing, linguistic, and other similar errors from the published text of a **normative document**.

Note: The results of correction may be presented, as appropriate by issuing a separate correction sheet or a **new edition** of the normative document.

Guide 2	Commentary

9.6 Amendment

Modification, addition, or deletion of specific parts of the content of a **normative document**.

Note: The results of amendment are usually presented by issuing a separate amendment sheet to the normative document.

9.7 Revision

Introduction of all necessary changes to the substance and presentation of a **normative document**.

In American National Standards (ANSI) parlance, a revision that takes place late (after ten years) is called a "new standard."

Note: The results of revision are presented by issuing a **new edition** of the normative document.

9.8 Reprint

New impression of a **normative document** without changes.

9.9 New Edition

New impression of a **normative document** that includes changes to the previous edition.

Note: Even if only the content of existing **correction** or **amendment** sheets is incorporated into the text of a normative document, the new text constitutes a new edition.

10. Implementation of Normative Documents

Note: A normative document can be said to be "implemented" in two different ways. It may be applied in production, trade, etc., and it may be taken over, wholly or in part, in another normative document. Through the medium of this second document, it may then be applied, or it may again be taken over in yet another normative document.

Guide 2	Commentary

10.1 Taking Over an International Standard (in a National Normative Document)

Publication of a national **normative document** based on a relevant **international standard,** or endorsement of the international standard as having the same status as a national normative document, with any deviations from the international standard identified.

Note: The term "adoption" is sometimes used to cover the same concept as "taking over," for example, "adoption of an international standard in a **national standard.**"

Adoption is frequently used in the United States.

10.2 Application of a Normative Document

Use of a **normative document** in production, trade, etc.

10.2.1 Direct Application of an International Standard

Application of an **international standard** irrespective of the **taking over** of that international standard in any other **normative document.**

10.2.2 Indirect Application of an International Standard

Application of an **international standard** through the medium of another **normative document** in which it has been taken over.

11. References to Standards in Regulations

11.1 Reference to Standards (in Regulations)

Reference to one or more **standards** in place of detailed **provisions** within a **regulation.**

In the United States, this is frequently called "incorporation by reference," especially in the regulations in the **Code of Federal Regulations (CFR).**

Note 1: A reference to standards is either dated, undated, or general, and at the same time either exclusive or indicative.

Note 2: A reference to standards may be linked to a more general legal provision referring to the **state of the art** or **acknowledged rules of technology**. Such a provision may also stand alone.

Guide 2	Commentary

11.2   Precision of Reference

11.2.1   Dated Reference (to Standards)

Reference to standards that identifies one or more specific **standards** in such a way that later **revisions** of the standard or standards are not to be applied unless the regulation is modified.

Note: The standard is usually identified by its number and either date or edition. The title may also be given.

11.2.2   Undated Reference (to Standards)

Reference to **standards** that identifies one or more specific **standards** in such a way that later **revisions** of the standard or standards are to be applied without the need to modify the **regulation**.

Because of the loss of control that this produces, most regulatory authorities do not use undated references.

Note: The standard is usually identified only by its number. The title may also be given.

11.3   Strength of Reference

11.3.1   Exclusive Reference (to Standards)

Reference to standards that states that the only way to meet the relevant **requirements** of a **technical regulation** is to comply with the **standard(s)** referred to.

11.3.2   Indicative Reference (to Standards)

Reference to standards that states that one way to meet the relevant **requirements** of a **technical regulation** is to comply with the **standard(s)** referred to.

Note: An indicative reference to standards is a form of a **deemed-to-satisfy provision**.

11.4   Mandatory Standard

Standard the application of which is made compulsory by virtue of a general law or **exclusive reference** in a **regulation**.

In the United States, "voluntary standards" frequently become mandatory when referenced in regulations.

12.   Conformity Assessment

The terms and definitions given in ISO/IEC 17000 apply.

2

ISO/IEC 17000 Conformity Assessment Vocabulary and General Principles

Introduction

Modern editions (sixth and seventh) of ISO/IEC Guide 2 included a core vocabulary for Conformity Assessment in clauses 12 through 17. The core vocabulary was prepared by the ISO Committee on Conformity Assessment, ISO/CASCO. In 2000, CASCO decided to remove the conformity assessment clauses 12 to 17 from Guide 2 and include the terminology for Conformity Assessment in a new (2004) International Standard, ISO/IEC 17000 Conformity Assessment—Vocabulary and general principles. Some of the Conformity Assessment terms in Guide 2 were retained in IS 17000, some were modified, and some new terminology and principles were included. To retain the connection of Guide 2 and IS 17000, a normative reference in Guide 2 is to IS 17000 and Guide 2 is cited "for related terminology" in IS 17000. The widespread use of the term "conformity," in a standards sense, is of relatively recent origin. It does not appear in IEC (1983) or ASTM (1990). ISO 9000:2000 disparages the use of "conformance." But that usage is deeply imbedded in information technology and telecom standards.

One of the striking differences in the new IS 17000, compared with the earlier Guide 2 editions, is that no definition of "conformity" is given. Only the term "Conformity Assessment" is defined in IS 17000. "Conformity" was defined in earlier Guide 2 editions as: "fulfilment by a product, process or service of specified **requirements**." Note that the definition of "requirements" has not changed.

Some of the Conformity Assessment terms in the seventh edition of Guide 2 were omitted from IS 17000. Conspicuous by their absence are the term "conformity evaluation" (14.1), and some dependent terms, defined by 14.1. The definition of conformity is rather intuitive and straightforward in many situations. But the meaning of conformity (to a standard) was called into question in more complex situations, such as computer programming language standards, at least as early as 1983 in ECMA (European Computer Manufacturers Association) Report TR/18, "The Meaning of Conformance to

Standards." T/R 18 was formally published in (ECMA 1983). Similar difficulties with defining conformity have persisted (Arnold 1994). The old definition of conformity is still useful in many areas, for example, ISO 9000 (Quality) International Standards. The new definition of conformity assessment in IS 17000 covers more territory, "specified requirements," which includes not only standards but technical specifications and regulations.

As in Chapter 1, the terms and definitions in IS ISO/IEC 17000 are reproduced in the following along with some commentary from a U.S. perspective. The IS 17000 Index is included in Appendix 2. Abbreviations and acronyms are collected in Appendix 3 for ease of reference.

ISO/IEC 17000	**Commentary**

1. Scope

This International Standard specifies general terms and definitions relating to conformity assessment, including the accreditation of conformity assessment bodies, and to the use of conformity assessment to facilitate trade.

A description of the functional approach to conformity assessment is included in Annex A, as a further aid to understanding among users of conformity assessment, conformity assessment bodies and their accreditation bodies in both voluntary and regulatory environments.

This International Standard does not set out to provide a vocabulary for all of the concepts that may need to be used in describing particular conformity assessment activities. Terms and definitions are given only where the concept defined would not be understandable from the general use of the term, or where an existing standard definition is not applicable.

Note 1: The notes appended to certain definitions offer clarification or examples to facilitate understanding of the concepts described. In certain cases, the notes may differ in different languages for linguistic reasons, or additional notes may be given.

Note 2: The terms and definitions are laid out in a systematic order, with an alphabetical index. A term in a definition or note that is defined in another entry is indicated by boldface, followed by its entry number in parentheses. Such terms may be replaced by their complete definition.

The alphabetical index appears in Appendix 2 of this book.

ISO/IEC 17000	Commentary

2. Terms relating to Conformity Assessment in General

2.1 Conformity Assessment

Demonstration that **specified requirements** (3.1) relating to a **product** (3.3), process, system, person or body are fulfilled.

Note 1: The subject field of conformity assessment includes activities defined elsewhere in this International Standard such as **testing** (4.2), **inspection** (4.3), and **certification** (5.5), as well as the **accreditation** (5.6) of **conformity assessment bodies** (2.5).

Note 2: The expression "object of conformity assessment" or "object" is used in this International Standard to encompass any particular material, product, installation, process, system, person, or body to which conformity assessment is applied. A service is covered by the definition of a product (see Note 1 to 3.3).

2.2 First-Party Conformity Assessment Activity

Conformity assessment activity that is performed by the person or organization that provides the object.

Note: The first-, second- and third-party descriptors used to characterize conformity assessment activities with respect to a given object are not to be confused with the legal identification of the relevant parties to a contract.

2.3 Second-Party Conformity Assessment Activity

Conformity assessment activity that is performed by a person or organization that has a user interest in the object.

Note 1: Persons or organizations performing second-party conformity assessment activities include, for example, purchasers or users of products, or potential customers seeking to rely on a supplier's management system, or organizations representing those interests.

Note 2: See Note to 2.2.

2.4 Third-Party Conformity Assessment Activity

Conformity assessment activity that is performed by a person or body that is independent of the person or organization that provides the object, and of the user interests in that object.

ISO/IEC 17000	Commentary
Note 1: Criteria for the independence of conformity assessment bodies and accreditation bodies are provided in the International Standards and Guides applicable to their activities (see Bibliography).	
Note 2: See Note to 2.2.	

2.5   Conformity Assessment Body

Body that performs conformity assessment services.	Underwriters Laboratories (UL) is an example.
Note: An **accreditation body** (2.6) is not a conformity assessment body.	

2.6   Accreditation Body

Authoritative body that performs **accreditation** (5.6).	In the United States, the National Voluntary Laboratory Accreditation Program (NVLAP) in NIST is an example.
Note: The authority of an accreditation body is generally derived from government.	

2.7   Conformity Assessment System

Rules, **procedures** (3.2) and management for carrying out **conformity assessment** (2.1).

Note: Conformity assessment systems may be operated at international, regional, national or subnational level.

2.8   Conformity Assessment Scheme

Conformity Assessment Programme

Conformity assessment system (2.7) related to specified objects of conformity assessment, to which the same **specified requirements** (3.1), specific rules and **procedures** (3.2) apply.	In the IEC, there are 3 conformity assessment schemes or systems (see Chapter 6).
Note: Conformity assessment schemes may be operated at international, regional, national or subnational level.	

2.9   Access

Access to a System or Scheme

Opportunity for an applicant to obtain **conformity assessment** (2.1) under the rules of the system or scheme.

ISO/IEC 17000	Commentary

2.10   Participant

Participant in a System or Scheme

Body that operates under the applicable rules without having the opportunity to take part in the management of the system or scheme.

2.11   Member

Member of a System or Scheme

Body that operates under the applicable rules and has the opportunity to take part in the management of the system or scheme.

3.   Basic Terms

3.1   Specified Requirements

Need or expectation that is stated.

Note: Specified requirements may be stated in normative documents such as regulations, standards and technical specifications.

Examples in the United States are FCC Regulations, UL Standards, and Publicly Available Specifications for software such as Sun's JAVA.

3.2   Procedure

Specified way to carry out an activity or a process.

[ISO 9000:2000, 3.4.5]

3.3   Product

Result of a process.

[ISO 9000:2000, 3.4.2]

Note 1: Four generic product categories are noted in ISO 9000:2000: services (e.g., transport); software (e.g., computer program, dictionary); hardware (e.g., engine mechanical part); processed materials (e.g., lubricant). Many products comprise elements belonging to different product categories. Whether the product is then called service, software, hardware or processed material depends on the dominant element.

Note 2: The statement of conformity described in Note 1 of 5.2 can be regarded as a product of **attestation** (5.2).

4.   Conformity Assessment Terms Relating to Selection and Determination (see Figure A.1)

4.1   Sampling

Provision of a sample of the object of conformity assessment, according to a **procedure** (3.2).

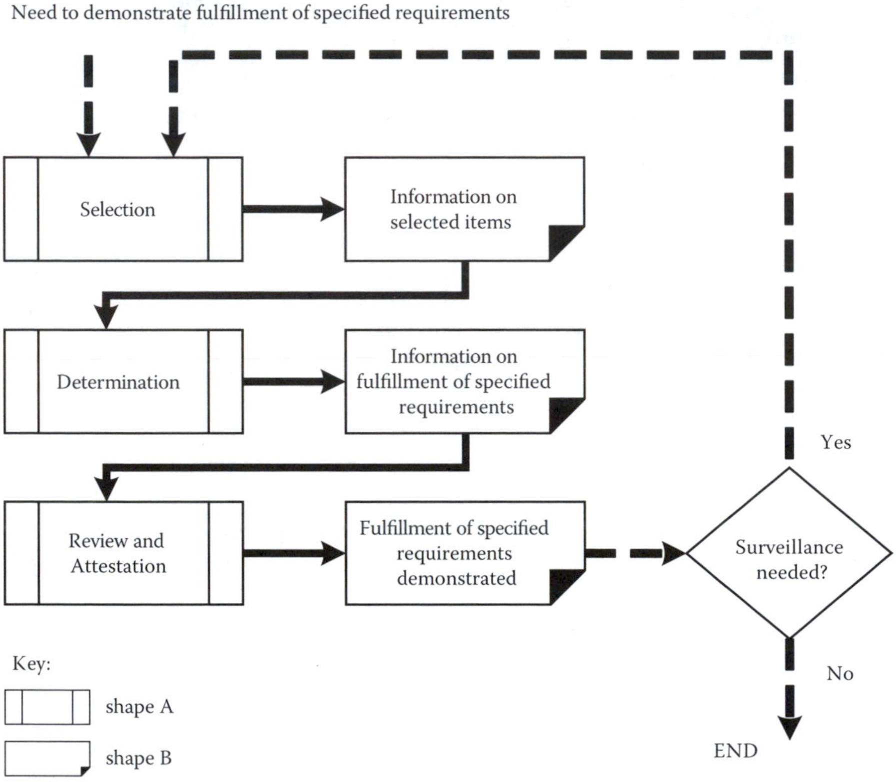

Need to demonstrate fulfillment of specified requirements

Key:

⬛ shape A

⬛ shape B

FIGURE A.1
Functional approach to conformity assessment.

ISO/IEC 17000	Commentary
4.2 Testing	
Determination of one or more characteristics of an object of conformity assessment, according to a **procedure** (3.2).	An older term "type testing" (14.5 in the seventh edition of Guide 2) is still used.
Note: "Testing" typically applies to materials, products or processes.	
4.3 Inspection	
Examination of a product design, **product** (3.3) process or installation and determination of its conformity with specific requirements or, on the basis of professional judgement, with general requirements.	
Note: Inspection of a process may include inspection of persons, facilities, technology and methodology.	

ISO/IEC 17000	Commentary

4.4 Audit

Systematic, independent, documented process for obtaining records, statements of fact or other relevant information and assessing them objectively to determine the extent to which **specified requirements** (3.1) are fulfilled.

Note: Whilst "audit" applies to management systems, "assessment" applies to conformity assessment bodies as well as more generally.

4.5 Peer Assessment

Assessment of a body against **specified requirements** (3.1) by representatives of other bodies in, or candidates for, an **agreement group** (7.10).

5. Conformity assessment terms relating to review and attestation (see Figure A.1)

5.1 Review

Verification of the suitability, adequacy and effectiveness of selection and determination activities, and the results of these activities, with regard to fulfillment of **specified requirements** (3.1) by an object of conformity assessment.

5.2 Attestation

Issue of a statement, based on a decision following **review** (5.1), that fulfillment of **specified requirements** (3.1) has been demonstrated.

This new term (attestation) is not greatly different from the meaning of the term in law. See, for example, Black's Law Dictionary.

Note 1: The resulting statement, referred to in this International Standard as a "statement of conformity" conveys the assurance that the specified requirements have been fulfilled. Such an assurance does not, of itself, afford contractual or other legal guarantees.

Note 2: First-party and third-party attestation activities are distinguished by the terms 5.4 to 5.6. For second-party attestation, no special term is available.

5.3 Scope of Attestation

Range or characteristics of objects of conformity assessment covered by **attestation** (5.2).

ISO/IEC 17000	Commentary

5.4   Declaration

First-party **attestation** (5.2).

5.5   Certification

Third-party **attestation** (5.2) related to products, processes, systems or persons.

The crucial element here is *third-party* (emphasis added). U.S. examples are Listing, Recognition, etc., for UL. CSA uses Certification consistently and some European Agencies use "approval."

Note 1: Certification of a management system is also sometimes called registration.

Note 2: Certification is applicable to all objects of conformity assessment except for **conformity assessment bodies** (2.5) themselves, to which accreditation is applicable.

5.6   Accreditation

Third-party **attestation** (5.2) related to a **conformity**.

Assessment body (2.5) conveying formal demonstration of its competence to carry out specific conformity assessment tasks.

6.   Conformity assessment terms relating to surveillance (see Figure A.1)

6.1   Surveillance

Systematic iteration of conformity assessment activities as a basis for maintaining the validity of the statement of conformity.

6.2   Suspension

Temporary invalidation of the statement of conformity for all or part of the specified scope of **attestation** (5.3).

6.3   Withdrawal

Revocation

Cancellation of the statement of conformity.

6.4   Appeal

Request by the provider of the object of conformity assessment to the **conformity assessment body** (2.5) or **accreditation body** (2.6) for reconsideration by that body of a decision it has made relating to that object.

ISO/IEC 17000	Commentary

6.5 Complaint

Expression of dissatisfaction, other than **appeal** (6.4), by any person or organization to a **conformity assessment body** (2.5) or **accreditation body** (2.6), relating to the activities of that body, where a response is expected.

7. Terms relating to conformity assessment and facilitation of trade

7.1 Approval

Permission for a **product** (3.3) or process to be marketed or used for stated purposes or under stated conditions.

Note: Approval can be based on fulfillment of **specified requirements** (3.1) or completion of specified **procedures** (3.2).

Commentary: Many European product safety agencies such as VDE (Verband Deutscher Elektrotechniker in Germany) still call their Certifications "Approvals."

7.2 Designation

Government authorization of a **conformity assessment body** (2.5) to perform specified conformity assessment activities.

7.3 Designating Authority

Body established within government or empowered by government to designate **conformity assessment bodies** (2.5), suspend or withdraw their designation or remove their suspension from **designation** (7.2).

Commentary: In the United States, the Department of Labor is the designating authority for Nationally Recognized Testing Laboratories (NRTLs).

7.4 Equivalence

Equivalence of conformity assessment results.

Sufficiency of different conformity assessment results to provide the same level of assurance of conformity with regard to the same **specified requirements** (3.1).

Commentary: In UL parlance, this is called an "equivalency clause" and is boilerplate in most UL standards.

7.5 Recognition

Recognition of conformity assessment results.

Acknowledgement of the validity of a conformity assessment result provided by another person or body.

Commentary: Not to be confused with a UL Certification (Recognition) for components.

7.6 Acceptance

Acceptance of conformity assessment results

Use of a conformity assessment result provided by another person or body.

ISO/IEC 17000	Commentary

7.7 Unilateral Arrangement

Arrangement whereby one party recognizes or accepts the conformity assessment results of another party.

7.8 Bilateral Arrangement

Arrangement whereby two parties recognize or accept each other's conformity assessment results.

7.9 Multilateral Arrangement

Arrangement whereby more than two parties recognize or accept one another's conformity assessment results.

7.10 Agreement Group

Bodies that are signatories to the agreement on which an arrangement is based.

7.11 Reciprocity

Relationship between two parties where both have the same rights and obligations toward each other.

Note 1: Reciprocity can exist within a multilateral arrangement comprising a network of bilateral reciprocal relationships.

Note 2: Although rights and obligations are the same, opportunities emanating from them can differ, this can lead to unequal relationships between parties.

7.12 Equal Treatment

Treatment accorded to **products** (3.3) or processes from one supplier that is no less favourable than that accorded to like products or processes from any other supplier, in a comparable situation.

7.13 National Treatment

Treatment accorded to **products** (3.3) or processes originating in other countries that is no less favourable than that accorded to like products or processes of national origin in a comparable situation.

ISO/IEC 17000	Commentary

7.14 Equal and National Treatment

Treatment accorded to **products** (3.3) or processes originating in other countries that is no less favourable than that accorded to like products or processes of national origin, or originating in any other country, in a comparable situation.

In the following informative Annex A of IS 17000, the aspects of actual conformity assessment, as practiced by Underwriters Laboratories (UL), are used as an illustrative example.

Annex A

(informative)

Principles of conformity assessment	Commentary

A.1 Functional approach

A.1.1 Conformity assessment is a series of three functions that satisfy a need or demand for demonstration that specified requirements are fulfilled:

– selection,
– determination; and
– review and attestation.

Such demonstration can add substance or credibility to claims that specified requirements are fulfilled, giving users greater confidence in such claims. Standards are often used as the specified requirements since they represent a broad consensus of what is wanted in a given situation. As a result, conformity assessment is often viewed as a standards-related activity.

A.1.2 Conformity assessment may be applied to products (defined to include services), processes, systems and persons, and also to those bodies that perform conformity assessment services. For convenience within this International Standard, the expression "object of conformity assessment" is used to refer collectively to any or all of these entities.

Annex A (continued)

(informative)

Principles of conformity assessment	Commentary

A.1.5   Shape B in Figure A.1 represents output from a function and is also the input to the next function. The nature of the output varies, depending on the specific activities that have been undertaken.

A.1.6   The solid arrows in Figure A.1 link the conformity assessment functions and their outputs/inputs. The broken arrows express the possible needs or demands for conformity assessment.

A.1.7   Conformity assessment activities can be characterized as "first-party," "second-party," or "third-party." Generally, for each of these categories

– the conformity assessment activities are under the control or direction of the type of individual or body stated in the definition, and

– the critical decision on which attestation is based is made by the type of individual or body stated in the definition.

A.2   Selection

A.2.1   Selection involves planning and preparation activities in order to collect or produce all the information and input needed for the subsequent determination function. Selection activities vary widely in number and complexity. In some instances, very little selection activity may be needed.

In a UL Investigation, the applicable standard (or standards) is usually known. If not, an appropriate ad hoc standard can be developed for a totally new product.

A.2.2   Some consideration may need to be given to selection of the object of conformity assessment. Frequently, the object may be a large number of identical items; ongoing production, a continuous process or a system; or involve numerous locations. In such cases, consideration may need to be given to sampling, or selection of specimens to be used for determination activities. For example, the sampling plan for river water, related to a demonstration that pollution requirements are fulfilled, would be an example of a sizeable and significant sampling activity. However, occasionally the object may be the whole population, for instance when a single, individual product is the object of conformity assessment. Even in such cases, sampling may be necessary to select a part of the entire object that is representative of the whole (e.g., selection of critical parts of a bridge for a determination of material fatigue).

Principles of conformity assessment	Commentary
A.2.5 Finally, additional information may be needed in order to perform determination activities properly so that the demonstration that specified requirements are fulfilled will be effective. For example, the scope of testing to be covered by laboratory accreditation must be identified before appropriate determination activities can be performed. Or, a description of a service may be needed before performing appropriate determination activities. Also, a determination activity may be a review of information alone, and that information must be identified and collected. For example, a copy of a product's instructions for use or warning markings may be needed.	This is usually a UL requirement for a Listed (complete) equipment.
A.2.6 In Figure A.1, all the information, samples (if sampling is used), decisions and other output from the selection function is represented as "information on selected items."	
A.3 Determination	
A.3.1 Determination activities are undertaken to develop complete information regarding fulfillment of the specified requirements by the object of conformity assessment or its sample. Some types of determination activities are defined in Clause 4.	UL calls this an Investigation. Inspection; Testing and Documentation are required.
A.3.2 The terms **testing** (4.2), **inspection** (4.3), **audit** (4.4) and **peer assessment** (4.5), which are defined as types of determination activities only, may used with "system" or "scheme" to describe conformity assessment systems or schemes that include the type of determination activity indicated. Thus a "peer assessment system" is a conformity assessment system that includes peer assessment as the determination activity.	
A.3.3 Various determination activities have no specific name or designation. An example is the examination or analysis of a design, or other descriptive information, in relation to specified requirements. Individual sub-fields of conformity assessment (e.g., testing, certification, accreditation) may have terms defined for determination activities that are unique to that sub-field. There is no generic term used in this International Standard or in practice to represent all determination activities.	There was in the seventh edition of Guide 2. It was 14.1, Conformity Evaluation.

Annex A (continued)

(informative)

Principles of conformity assessment	Commentary

A.4   Review and Attestation

A.4.1   Review (5.1) constitutes the final stage of checking before taking the important decision as to whether or not the object of conformity assessment has been reliably demonstrated to fulfill the specified requirements. Attestation (5.2) results in a "statement" in a form that most readily reaches all of the potential users. "Statement of Conformity" is a generic expression used to include all means of communication that fulfillment of specified requirements has been demonstrated.

> If the UL engineer doing the investigation concludes that specified requirements are fulfilled, the report is given to a Reviewer. If that is positive, a Listing Report (attestation) is issued along with a Follow-Up Service Procedure.

A.4.2   If fulfillment of the specified requirements has not been demonstrated, the finding of nonconformity may be reported.

> UL calls that a "letter report."

A.4.3   The terms **declaration** (5.4), **certification** (5.5) and **accreditation** (5.6), which are defined as types of **attestation** only, may be used with "system" or "scheme" to describe conformity assessment systems or schemes that include the type of attestation activity indicated as the final step. Thus, a "certification system" is a conformity assessment system that includes selection, determination, review and finally certification as the attestation activity.

A.4.4   In Figure A.1, all the output from the review and attestation function is represented as "fulfillment of specified requirements demonstrated."

A.5   Need for surveillance

A.5.1   Conformity assessment can end when attestation is performed. However, in some cases systematic iteration of the functions in Figure A.1 may be needed to maintain the validity of the statement resulting from attestation. The needs of users drive such activities. For example, an object of conformity assessment may change over time, which could affect its continuing fulfillment of specified requirements. Or, users may demand ongoing demonstration that specified requirements are fulfilled; for example, when a product is produced continually.

> The National Electrical Code (NEC) requires that Nationally Recognized Testing Laboratories conduct periodic inspection of production of Listed (Certified) equipment.

Principles of conformity assessment	Commentary
A.5.4 Choices about the specified requirements can be different as well. For example, only a subset if the specified requirements might be selected in any given iteration of surveillance. Or, similarly, only a portion of the object of conformity assessment may be selected for determination activities in surveillance; for example, only a portion of an accredited certification body may be audited during surveillance.	The Specified Requirements for UL surveillance are defined in the Follow-Up Service Procedure.
A.5.5 As noted above, the different choices in selection can lead to different determination activities for surveillance purposes. However, in both initial assessment and surveillance, the output from selection defines the determination activities and how they will be carried out.	
A.5.6 The review and attestation function is also used in both initial assessment and surveillance. In surveillance, a review of all the inputs and outputs in Figure A.1 leads to a decision whether the statement resulting from attestation continues to be valid. In many cases, no special action is taken if the statement continues to be valid. In other cases, for example, if the scope of attestation has been extended, a new statement of conformity might be issued.	
A.5.7 If the decision is that the statement of conformity is no longer valid, appropriate activities are necessary to advise users; for example, that the scope of the attestation has been reduced or that the statement has been suspended or withdrawn.	UL calls that a Variation Notice (VN). Labeled production may be stopped until the VN is cleared.

Annex B

(informative)

Related terms defined in other documents

B.1 General terms given specific meanings in the ISO/IEC 17000 series

B.1.1 Compound terms defined for the purposes of ISO/IEC 17011, relating to accreditation

accreditation body logo

accreditation certificate

accreditation symbol

extending accreditation

Annex B (continued)

(informative)

Related terms defined in other documents

reducing accreditation

scope of accreditation

suspending accreditation

B.1.2 General terms given specific meanings for the purposes of ISO/IEC 17024, relating to certification of persons

competence

evaluation

examiner

qualification

B.1.3 Compound terms defined for the purposes of ISO/IEC 17030, relating to marks of conformity

issuer of a third-party mark of conformity

owner of a third-party mark of conformity

third-party mark of conformity

B.2 Terms defined in standards outside the ISO/IEC 17000 series

calibration	VIM
capability	ISO 9000
characteristic	ISO 9000
competence	ISO 9000
conformity	ISO 9000
customer	ISO 9000
document	ISO 9000
information	ISO 9000
measurement	VIM
organization	ISO 9000
process	ISO 9000
specification	ISO 9000
supplier	ISO 9000
system	ISO 9000
validation	ISO 9000
verification	ISO 9000

The Bibliography for ISO/IEC 17000 groups applicable standards (in the ISO/IEC 17000 series) by subject matter.

Bibliography

For conformity assessment systems based on determination activity

[1] ISO/IEC 17020:1998, *General Criteria for the operation of various types of bodies performing inspection*

[2] ISO/IEC 17025:1999, *General requirements for the competence of testing and calibration laboratories.*

[3] ISO/IEC 17040:2005, *Conformity assessment—General requirements for peer assessment of conformity assessment bodies and accreditation bodies*

[4] ISO 19011:2002, *Guidelines for quality and/or environmental management systems auditing*

For conformity assessment systems based on attestation activity

[5] ISO/IEC 17011:2004, *Conformity assessment—General requirements for accreditation bodies accrediting conformity assessment bodies*

[6] ISO/IEC 17021:2006, *Conformity assessment— Requirements for bodies providing audit and certification of management systems*

[7] ISO/IEC 17024:2003, *Conformity assessment—General requirements for bodies operating certification of persons*

[8] ISO/IEC Guide 65:1996, *General requirements for bodies operating certification systems*

For statements of conformity

[9] ISO/IEC 17030:2003, *Conformity assessment—General requirements for third-party marks of Conformity*

[10] ISO/IEC 17050-1:2004, *Conformity assessment—Supplier's declaration of conformity*—Part 1: General requirements

For recognition and acceptance of conformity assessment results

Commenttary

A second edition was issued in 2005 (05-15).

[5] cancels and replaces ISO/IEC Guides 58, 61 and ISO/IEC/TR 17010.

Note: This IS replaced Guides 62 and 66. They are still in the list of this chapter for completeness and history.

Note: ISO/IEC 17050-2:2004, *Conformity assessment – supplier's declaration of conformity*—Part 2: Supporting document had not been issued when IS 17011 was published.

[11] ISO/IEC Guide 68:2002, *Arrangements for the recognition and acceptance of conformity assessment results*

For related terminology

[12] ISO 3534 (all parts), *Statistics—Vocabulary and symbols*

[13] ISO 9000:2000, *Quality management systems— Fundamentals and vocabulary*

[14] ISO/IEC Guide 2:2004, *Standardization and related activities—General vocabulary*

[15] VIM (1993), *International vocabulary of basic and general terms in metrology*, BIPM, IEC, IFCC, ISO, IUPAC, IUPAP and OIML

Other ISO/IEC Guides and Publicly Available Specifications produced by ISO/CASCO:

ISO/IEC Guide 7:1994, *Guidelines for drafting of standards suitable for use for conformity*

ISO/IEC Guide 23:1982, *Methods of indicating conformity with standards for third-party certification system*

ISO Guide 27:1983, *Guidelines for corrective action to be taken by a certification body in the event of misuse of its mark of conformity*

ISO/IEC Guide 28:2004, *Conformity assessment—Guidance, on a third party certification system for products*

ISO/IEC Guide 43-1:1997, *Proficiency testing by interlaboratory comparisons—Part 1 Development and operation of proficiency testing schemes*

ISO/IEC Guide 43-2:1997, *Proficiency testing by interlaboratory Comparisons—Part 2 Selection and use of proficiency tests by laboratory accreditation bodies*

ISO/IEC Guide 53:2005, *Conformity assessment—Guidance on the use of an organization's quality management system for certification*

ISO/IEC Guide 60:2004, *Conformity assessment—Code of good practice*

ISO/IEC Guide 62:1996*, *General requirements for bodies operating assessment and certification/Registration of quality systems*

ISO/IEC Guide 65:1996, *General requirements for bodies operating product certification systems*

ISO/IEC Guide 66:1999*, *General requirements for bodies operating product certification systems*

ISO/IEC Guide 67:2004, *Conformity assessment— Fundamentals of product certification*

ISO/IEC Guide 68:2002, *Arrangements for the recognition and acceptance of conformity assessment results*

ISO/PAS 17001:2005,*Conformity assessment—Impartiality— Principles and requirements*

ISO/PAS 17002:2004, *Conformity assessment— Confidentiality—Principles and requirements*

ISO/PAS 17003:2004, *Conformity assessment—Complaints and appeals—Principles and requirements*

ISO/PAS 17004:2005, *Conformity assessment—Disclosure of information—Principles and requirements*

*These ISO/IEC Guides were replaced in 2006 by a standard: ISO/IEC 17021.

Note: Guides and Standards are constantly being revised or replaced. Check listings on CASCO Web sites (links on www.iso.org) and catalog fo∫r latest information.

References

Arnold, David, 1994, Conformance Testing ICT Standards, Costly Chore or Valuable Insurance? *Association for Computing Machinery, Standard View* 2(2): 182–187.

ASTM, 2000, *ASTM Dictionary of Engineering Science & Technology*, 10th ed., West Conshohocken, PA, ASTM International.

ECMA, 1983, The Meaning of Conformance to Standards (TR/18), *Computer Standards and Interfaces*, pp. 201–207.

IEC, 1983, *IEC Multilingual Dictionary of Electricity*, New York, Wiley Interscience.

3

Standards

The Economic Benefits of Standardization

To those familiar with standardization, the economic benefits seem to be (and are) self-evident. But standardization is expensive and it is frequently necessary to show that it is economically beneficial, both at the micro- and macroeconomic levels. At the microeconomic level, a demonstration of cost savings due to variety reduction can be used to demonstrate the benefits. Good work in this area can be traced back to Toth (1984) and earlier. These subjects were common topics of several annual conferences of the Standards Engineering Society (SES) and in its journal *Standards Engineering*. Variety reduction is, of course, not the only way that standardization reduces costs, but it is an important aspect. If Certification of a standard part is also used, it can reduce the costs of configuration management as outlined by Hunter (1992). The failure of a producer to comply with standards and regulations expected by the consumer can result in a loss of markets or worse (lawsuits).

It is considerably more difficult to show the economic benefits of standardization at the macroeconomic level, but recently published studies demonstrate the benefits and the costs of standardization at both national and regional levels. These studies quantify the extent to which standardization can encourage or impede innovation. This discussion continues in Chapter 12, "Management Aspects of Standards, Conformity Assessment, and Accreditation." Standardization starts, of course, with a standard, so that is the next subject.

Anatomy of a Standard

The format or *presentation* of various standards varies from one Standards Body, or Standards Developing Organization (SDO), to another and even from one type of standard to another coming from the same Standards Body.

Standards do, however, tend to have a certain anatomy or presentation that is outlined next with an example of an International Standard.

Cover: The cover of a standard normally includes the name and logo of the SDO, the number and title of the standard, and the date (at least approximately) of its publication. For IEC standards, the identification of the publication is in French (Norme Internationale) as well as English (International Standard). If the standard is Part 1: General rules, it also has a "-1" in its number and generally applies to all the Particular Requirements of the following dash number parts of the standard.

Boilerplate: The term "boilerplate" is not, of course, recognized standards terminology, but the items found inside the cover of many standards are frequently preliminary statements, terminology, and references to other Basic Standards. A table of contents is normally provided in all but the simplest standards.

Foreword: The foreword generally includes some more boilerplates, as well as some history of the standard's development. Disclaimers on product marking (no Certification marking authorized by this standard) and on patent rights (possibility that patent rights may cover some aspects of the standard) are usually included. The foreword may explain how articulated type fonts are used to distinguish Requirements, Test specifications, and Notes.

Scope: The Subject of standardization (topic) and Field of standardization are spelled out in the scope statement to clarify the applicability of the standard. This important part of a standard is meant to assure that gaps and duplication do not occur in view of other standards in related areas. It is sometimes helpful to include a "This standard does not apply to:" statement to further clarify the boundaries of the territory covered by the standard.

Normative references: This part of a typical standard identifies Normative documents that, by being referred to in the standard, become a part of the standard. Because such Normative documents may be revised from time to time, Dated references will have to be reviewed after revision to determine how they are to be handled (if revised Normative documents of a later date are still applicable). In some situations it may be desirable to use Undated references so long as such Normative references do not materially change the scope and intent of the standard.

Terms and definitions: Special technical terms and their definitions (terminology) are a crucial part of every standard. As will be seen in Chapter 4, "Standards Development," the terms of a standard are drawn from specialized dictionaries or vocabularies and are also defined in this section of a standard when the more general vocabularies need supplementation. Special technical terms and their definitions specific to the standard must be crafted very carefully to avoid ambiguity.

Classification: In some standards, a classification scheme is used to distinguish the Subjects of standardization.

Characteristics: A summary of the characteristics of the items being standardized appears in relatively complex standards. This section will not appear in all standards.

Marking: For a component to be safely and intelligently applied, the standards may require certain (not Certification) marking. For example, the electrical ratings (voltage, current, etc.) of an electrical component are normally required. The details of many such markings are specified by other standards, for example, IEC 417M—Graphical symbols for use on equipment.

Standard conditions for use: An electrical component's use in an end equipment (or next higher assembly) is subject to limitations that include the ambient temperature, altitude, and relative humidity among others. These standard conditions for use are variously called installation instructions or, in Underwriters Laboratories (UL) standards, Conditions of Acceptability (CoAs).

Requirements: The Requirements (7.5 of Guide 2) convey the criteria that must be fulfilled for the object to satisfy the standard. Requirements are the body of many product standards. In other types of standards, this part of the standard is called an Instruction. Contrast these items with Recommendations, which only convey advice or guidance. Requirements are also frequently included in Normative annexes.

Tests: Electrical, mechanical, and other tests usually appear in product standards and obviously are the corpus of Test Standards. Type tests (the testing of a single piece of equipment or component) are usually given in product standards. Production line tests may also be included in a product or related standard.

Normative annexes: Normative annexes frequently appear in standards to identify requirements that must be met for the fulfillment of the standard. These are usually standards or other documents.

Informative annexes: Informative annexes are used in standards to convey advice, understanding, and other information that does not include a Requirement.

Bibliographies: Bibliographies are included in some standards to provide additional information in the form of books and papers on subjects germane to the subject of the standard.

Indexes: An index enhances the usability of a complex or lengthy standard.

Variations on a Theme

Some Standards Bodies prescribe several variations of the Standard. The American Society for Testing and Materials International (ASTMI), for example, has several formats for standards depending on the type of standard involved:

Format for Test Methods

Format for Specifications

Format for Other Types of Standards

A similar scheme is recommended by the Standards Engineering Society (SES) in its Recommended Practice for Designation and Organization of Standards (SES-1:2002).

Modularity in Standards

Modularity is a salient feature in architecture, computer hardware, computer programs, and many other fields. One form of that in standards is "dash-number modularity," which is frequently embodied in a mature set of IEC Standards: IEC 60335—Safety of household and similar electrical appliances. The first module of the set is: IEC 60335-1 (date) Part 1: General Requirements. This Part 1 includes, with some few exceptions, requirements for household electrical appliances that are common to a wide variety of appliances such as vacuum cleaners, electric irons, dishwashers, and so forth. Each of the subsequent dash-number parts of the standard, included in Part 2, is called Particular requirements, for example, Particular requirements for electric irons (60335-2-3). There are over two dozen additional dash-number parts with Particular requirements for other types of household electrical appliances. It thus becomes unnecessary to repeat the General requirements in each of a set of specialized standards and makes subsequent modifications simpler.

A second type of modularity in standards, mentioned earlier, is the Normative annex. The Normative annex lists additional standards (Provisions, frequently Requirements) that are as much a part of the standard as if they had been included in full. Dated references (11.2.1 of Guide 2) include dated standards and are to be used without revision (a later date or edition) unless the standard in which the reference is included is modified to account for another version of the reference. Undated references (11.2.2 of Guide 2) include undated standards with the understanding that later versions will automatically be "incorporated by reference" without modification of the standard in which the reference is included. That will require that the Standards body review the continued applicability of references, which, although undated, may have changed since the last revision. Undated references in ASTMI standards are preferred unless a compelling reason exists for the use of a dated reference. The policies of other standards developers vary.

Another kind of modularity in standards, that might be called "serial modularity," is often used in standards called *Profiles*, or *functional standards*, used especially in Information Communication Technology (ICT) standards. The building block standards of the Profile are concatenated with standard interfaces (see Figure 3.1). That permits some variation of the individual modules without destroying the interoperability of the scheme defined by the profile. Such modules are called *objects* in computer programs.

Descriptive Provisions versus Performance Provisions in Standards

A Descriptive provision (7.7 in Guide 2) is frequently called a "design standard" in the United States. As indicated in the Note of 7.7, Descriptive provisions frequently specify the dimensions and material of the object. A Performance provision (7.8 of Guide 2) on the other hand, frequently called a "performance standard" in the United States, focuses on the fitness for purpose or the behavior of the object. The term "performance standard" has, in the past, been confused with the overall performance (speed, etc.) of an object, such as a computer. Performance provisions are frequently preferred to design standards so that there is room to use new technology and for legal reasons, as will be seen in Chapter 10.

The Requirements in some standards are largely Descriptive and in other standards are largely Performance oriented. In many cases a standard

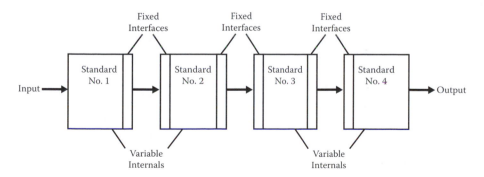

FIGURE 3.1
The serial modularity of standards in a profile.

includes a mix of these types of requirements. A well-known standard that is almost completely Descriptive is the standard for the ubiquitous modular plug and jack (Figures 3.2 and 3.3, respectively). The Federal Communications Commission (FCC) made the modular plug and jack a regulatory standard

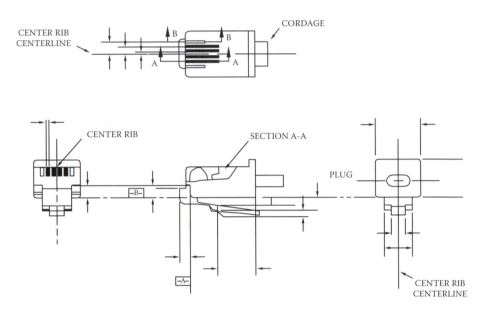

FIGURE 3.2
Simplified FCC specifications for the modular plug.

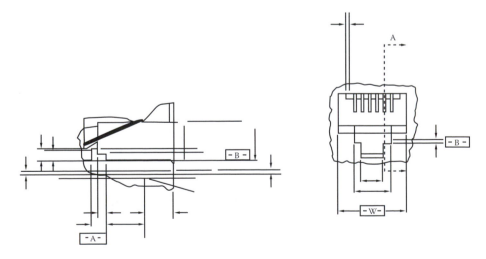

FIGURE 3.3
Simplified FCC specifications for the modular jack.

by publishing it in the Code of Federal Regulations (47 CFR, Part 68). Since the whole idea behind the modular plug and jack was interchangeability, the dimensions were carefully and precisely specified as was the contact material. The FCC wanted to leave the door open to improvements but included the following statement to assure continued interchangeability:

> The plug and jack designs shown are representative of generic types, and should not be interpreted as the only designs that may be used. Design innovation and improvement is expected; *but for interchangeability to be maintained, alternative designs ... must be compatible with the plugs and jacks shown. The interface dimensions between mating plugs and jacks must be maintained.* (emphasis added)

Note: The specifications for the modular plugs and jacks have now been privatized and the custodian of these standards is the Administrative Council for Terminal Attachments (ACTA) at www.part68.org.

As an example of a Performance provision (7.8 of Guide 2), the Comparative Tracking Index (CTI) for evaluating the resistance to tracking on insulation surfaces, for example, printed wiring boards (PWBs), is instructive. One of the ways insulation can fail is to develop a conducting track between conductors (e.g., traces) on the surface of the insulation on which the conductors are mounted. The ability of the PWB material to resist tracking is used to specify the minimum distance (creepage) between the traces for safety requirements.

A schematic representation of a test setup for measuring CTI is shown in Figure 3.4. Two electrodes are placed against the surface of the material being tested. A drip tank above the point between the electrodes is adjusted to drip a specified solution on the area between the electrodes at a certain rate. A

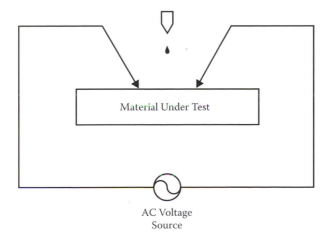

FIGURE 3.4
Simplified schematic of the CTI test setup.

voltage is applied between the electrodes and the number of drops it takes for a track to develop that results in a "short circuit" between the electrodes is noted. The greater the number of drops, the higher the CTI and the resistance to tracking. The result does not depend on the thickness of the material under test but depends only on its surface electrical properties. The International Standard for the CTI test is IEC 60112 *Method of determining the comparative and the proof tracking indices of solid insulating materials under moist conditions.*

Good Standards versus Bad Standards

Good Standards can be characterized in a number of ways. The following are the salient characteristics of a good Standard:

- It meets the socioeconomic, political, trade, and technical needs for which it is developed with a clear scope and purpose.
- It reflects the real needs of users as much as possible.
- It arrives in a timely fashion, not too soon and not too late.
- Its scope is appropriate, not too broad and not too narrow.
- It does not include more than several levels of abstraction.
- It is readily understood by people who did not help prepare it.
- It supplies all needed definitions internally or by reference.
- It encourages innovation by the use of performance requirements where possible.
- It uses state of the art provisions.
- It fits properly into a framework of other standards.
- It is structured to facilitate Conformity Assessment.
- All tests included have been validated.

Bad standards would, of course, fail to have these characteristics.

Other Types of and Names for Standards Not Identified in ISO/IEC Guide 2

Some of these "standards" meet the definition in Guide 2 and others do not. The reader of the literature on standards and conformity assessment will encounter these terms.

- De facto standards: These documented or undocumented standards are those that become predominant due to wide use. The internal company technical standards of the old AT&T organization provide good examples of such standards. Other frequently mentioned de facto standards include the Windows™ Operating System, the IBM Personal Computer, the QWERTY keyboard layout, and the TCP/IP suite of Internet protocols. A less frequently encountered de facto standard is the form, fit, and function of the Astec power supply, which was widely used in personal computers.

- De jure standards: Most modern writers call these consensus Standards (they normally meet the Guide 2 definition of a Standard). Other writers use the term de jure standard to mean a Regulation (as defined in Guide 2). The result is an ambiguity that points up the importance of using the ISO/IEC Guide 2 terminology to minimize confusion when discussing standards.

- Proprietary standards: These are "standards" developed within a single company or agency and do not meet the Guide 2 definition of a standard. Some proprietary "standards" are also de facto standards, for example, the Windows operating system.

- Open standards: Open standards are those developed in a public (rather than private) forum. A Publicly Available Standard (PAS) developed by a consortium would not be an example of an open standard.

- Other names used where consensus has failed or for other reasons include Deliverables (ISO), Technical Reports (IEC), and Recommended Practices (IEEE).

Codes and Other Pre-Regulation (Pre-Reg) Standards

Many consensus standards that fully meet the ISO/IEC definition are destined to become Codes or Regulations. Prime examples of such Pre-Reg standards are the National Fire Protection Association (NFPA) Codes and Standards such as the National Electrical Code (NEC) that are routinely included in Federal, State, and Local Regulations.

Thus, parts of our laws are prepared not by legislators directly but by panels (like working groups) of experts. Government officials may participate in the preparation of a Code or Standard but that is not necessary for their adoption. The proposed adoption of such standards is normally announced in the *Federal Register* or a state register, as appropriate. An outline of the process for NFPA Codes and Standards appears in a paper by Grant (1996).

Applications of Standards

In addition to standardization per se, standards may be used in a variety of ways. As an example, consider IEC Standard 61204, Low-voltage power supply devices, d.c. output—Performance standards. This standard, actually a family of dash-number standards, is intended to provide a convenient way of specifying the electrical characteristics of low-voltage power supplies. It includes, for example, ways to specify the "foldback characteristics" of a power supply (the way the output voltage folds back under overload). The safety requirements of power supplies are generally included in the end equipment standards rather than "component" standards since medical equipment has different safety requirements than computer equipment. An object of IEC 61204 was to provide "graded" performance levels to minimize the "specmanship" that permeates some power supply data sheets.

One could use this or a similar standard as a "checklist" to see if all the important characteristics of a power supply being designed or procured have been included in a company power supply specification. Another use of this or a similar specification is to incorporate it into a design or procurement specification by reference. This can speed up the preparation and approval of such specifications. Yet another use for a standard is in an Attestation or manufacturer's declaration. That is a marketing tool for broadening the sales of such units. Finally, a standard may be used as the basis for Attestation (including Certification and Accreditation), the highest uses to which a standard may be put. More on these issues appear in Chapter 5, "Accreditation of Conformity Assessment Bodies (CABs)—ISO/IEC 17011."

There is an international organization, the International Federation of Standards Users (IFAN). As indicated, this organization, which has the Standards Engineering Society as the American and Canadian representative, focuses on user needs in standards.

Abuses of Standards

There are several ways in which a standard may be abused. One is the deliberate misinterpretation of a standard so that conformity may be claimed. When a product fails to meet the requirements of a standard, the product may be improved so as to conform, or the standard may be downgraded to make a nonconforming product conforming. An example of the latter is described by F. Cecil Brenner (Spivak and Brenner 2001). Brenner describes how a Society of Automotive Engineers (SAE) standard for automobile tires was downgraded so that production tires could meet the standard. The results were disastrous. As a result, standards for tires were included in the Motor Vehicle and Highway Safety Acts of 1966. That forced a re-revision of the SAE standard for tires.

The Life Cycle of a Standard

Most standards go through a life cycle roughly like that described in the following and illustrated in Figure 3.5.

Stage 1:   The need for a standard must be recognized by a member (or members) of a standards body that can take action on it. Users should be involved whenever possible.

Stage 2:   The exact scope and principal provisions of the standard are specified.

Stage 3:   The detailed provisions of the standard are prepared and approved.

Stage 4:   The standard is tested or published for "preliminary use" to determine its suitability and user feedback is incorporated.

Stage 5:   The standard is formally published.

Stage 6:   The standard is applied (placed into use in *standardization*).

Stage 7:   After five years of use (or a similar period), the standard is reviewed to determine if it should be *revised, replaced, reaffirmed, or withdrawn.*

These stages will be discussed in more detail in Chapter 4, "Development of Standards," and in Chapter 6 as the stage codes for ISO and IEC standards.

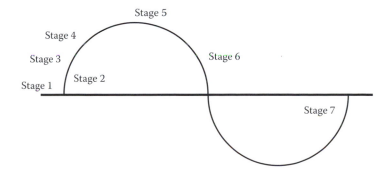

FIGURE 3.5
The stages in the life cycle of a standard.

References

Grant, Casey, 1996, The Making of an NFPA Code or Standard, *Standards Engineering* 48(6): 13–15.

Hunter, Robert D., 1992, Standards Enhance Competitiveness, *Standards Engineering* 44(6): 133–137.

Spivak, Steven M., and F. Cecil Brenner, 2001, *Standardization Essentials: Principles and Practice*, New York, Marcel Dekker, pp. 217–219.

Toth, Robert B., 1984, The Economics of Standardization—A Pragmatic Approach, in Robert B. Toth (Ed.), *The Economics of Standardization,* Minneapolis, Standards Engineering Society, pp. 1–44.

4

Standards Development

Introduction

One of the more interesting, and controversial, aspects of standards is in their development. The term "Standardization" is sometimes applied to the *process* of standards development but that term includes the *implementing* of standards as well as their development (see Note 1, par. 1.1 in Guide 2, Chapter 1). The development of standards is normally done by a Standards Body, a body that has recognized activities in this area. Such a body is frequently called a Standards Developing Organization (SDO) or a Standards Setting Organization (SSO). Internal company standards developments are too varied to be covered here, but a company should consider adopting (Taking over) International Standards if it has growth aspirations and plans to market its products abroad. More on that subject appears in Chapter 12, "Management Aspects of Standards, Conformity Assessment, and Accreditation."

The focus in this chapter is on three primary processes for developing standards:

1. Classical methods (also called traditional or formal methods)
2. Consortia and similar methods
3. Internet Engineering Task Force (IETF) methods

These methods should not be considered to be mutually exclusive. Many important standards development projects involve more than one of these approaches. It must be emphasized that the *sample* of standard development processes described in this chapter is just a sample, although a representative one, and that some other important processes are not described because of space limitations.

Classical Methods

The basic processes for the development of "voluntary consensus standards," as they were historically called, have changed somewhat over the years, but they still have features that are recognizable as being used since the turn of the century. The term *process* is stressed here. Process control, an engineering discipline in its own right, is a hallmark of this brand of standards development. A stable development process is essential to encourage implementation and to prevent disruptive revisions that come too soon (Schumny 2000). Standards must evolve but should not change so rapidly as to create instability. Process control is also espoused by Hohmann (1989).

The Emergence of "Due Process"

Two U.S. Supreme Court decisions in the 1980s had profound impacts on the classical method of standards development. The first, *American Society of Mechanical Engineers (ASME) v. Hydrolevel Corp.* in 1982, was actually a case involving Conformity Assessment (Interpretation), which was morphed into a standards development issue as well by the Federal Trade Commission (FTC). A major issue in *Hydrolevel* was the use of a "design requirement" rather than a "performance requirement" in the ASME standard involved. That issue added impetus to the ethic of using performance versus design requirements. ISO/IEC Guide 59 also has such a requirement (par. 5.6; ISO/IEC 1994).

The second Supreme Court decision, *Allied Tube and Conduit Corp. v. Indian Head, Inc.*, dealt more directly with standards development per se. The basic issue in that case was the "stacking" of a "standards ballot committee" to produce an egregious result. Both of these cases as well as other legal aspects of standards development and conformity assessment are discussed in Chapter 10 and are discussed by Swankin (1990). Subcommittee operation is described by Shanahan (2005) and balloting processes by Sheahan (2005).

It should be noted that "due process" in standards development does not have the same literal meaning as the term in law (e.g., Constitutional Amendments 5 and 14), although some would like to make it so. Due process in standards development involves the following features.

- Written procedures, usually supplied by the SDO, which guard against the creation of any requirements that are discriminatory and could restrain trade that would invite antitrust action against the SDO by the Federal Trade Commission (FTC) or the Department of Justice.
- Consensus: As defined in Guide 2, consensus does not imply unanimity but usually calls for more than a simple majority (typically two-thirds to three-fourths) favorable voting on the balloting of a

proposed standard after robust attempts have been made of resolve outstanding differences and objections.

- Openness: The process of developing a standard must be open to materially and directly affected parties (stakeholders). Requested information on the process must be provided in a timely fashion.

- Notice: Notification of standards development activities must be made in suitable fora, for example, ANSI Standards Action and elsewhere. For some ANSI Accredited Standards Developers (ASDs), a proposed standard that might become an ANSI standard must be published in Standards Action in the PINS (Project Initiation Notification System) pursuant to ANSI's Essential Requirements: Due Process Requirements for American National Standards. This permits materially and directly affected parties to request further information or submit comments. Proposed Foreign Government Regulations are also published in Standards Action, pursuant to WTO (World Trade Organization) requirements.

- Handling of comments and negative ballots: During the standards development process, the views of all stakeholders must be carefully considered and resolved. The handling of negative ballots is especially critical. Those casting negative votes on any ballot during the process must identify the reason for the objection so that the developer may consider revising a provision so as to convert a negative vote to a positive one. If the change is substantial, the draft standard must then be recirculated to assure that the change does not prompt other voters to change a previous yes vote to a no vote. This part of the process can be very circuitous! The detailed treatment of how to handle negative votes is, of course, prescribed by the rules of the SDO.

- Balance: To prevent a single interest from dominating a standards committee, as happened in the Allied Tube case, due process requires that the groups of members of such a committee must be apportioned according to the number of different kinds of stakeholders involved. For example, if there are three groups of stakeholders involved—producers, consumers and other—then *roughly* 33 percent of the total members should be included in each group.

- Record keeping: The keeping of good and complete records of all proceedings of a standards committee is an essential part of due process. This does not require a verbatim transcript of the proceedings but must record the major actions of the committee, especially the manner in which negative votes are resolved. These records should be available to any interested party at a reasonable cost.

- Use of patented items in a standard: Patented items should not be used in a standard unless there are no other alternatives and the patent owner is willing to license all comers at reasonable nondis-criminatory royalties. More on this subject appears in Chapter 10.

- Rationale statements: One of the best defenses against antitrust charges in the preparation of a standard, or its revision, is to include the rationale for each requirement. This shows the underlying motivation of each requirement and demonstrates that a requirement is not arbitrary or collusive.

- Appeal: Most modern SDO procedures include a requirement for an appropriate path of appeal on either procedural or technical issues. The "appeals board" of an SDO must meet certain impartiality requirements and be capable of implementing any of several actions, including the termination of the standard development. The right to appeal provision probably grew out of Government antitrust actions, but are not universally implemented in SDO Procedures. ANSI Accredited Standards Developers are required to have appeal procedures for procedural issues (actions or lack of actions). ANSI will not hear an appeal from an ASD unless the ASD's internal appeal process has been completed.

- Review and Maintenance: The Standards body must periodically review each issued standard at least every five years (or so, depending on its rules). The five-year rule apparently originated with ISO. Any standard of an ANSI accredited standard developer that is not revised, withdrawn, or reaffirmed within a ten-year period is administratively withdrawn by ANSI. The actions of the SDO must continue to comply with due process requirements. Some ten accredited standards developers are exempted from the requirement since they regularly review their standards at shorter intervals. The process is referred to as "Continuous Maintenance." The federal government itself is not, of course, required to follow such rules in its regulatory standards. This was pointed out by Penenberg (2003) in the matter of hopelessly outdated federal tire standards that contributed to the famous failures of Firestone tires on Ford Explorers.

Simplified Organization of a Standards Body (SDO)

Figure 4.1 shows a simplified organization of an SDO. The top block includes the part of the organization that sets overall policies and procedures, provides a "manual" for standards development, including templates and other automation support. It verifies that committee actions meet due process requirements and handles the final balloting of a proposed standard. If the results are positive, it issues the standard.

The next level (left-hand block) includes the detailed "off-line" administrative functions not directly concerned with the development of standards.

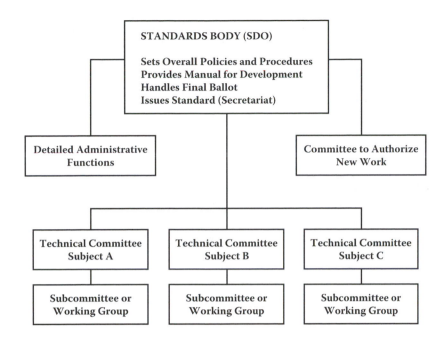

FIGURE 4.1
Simplified organization of a standards body.

The right-hand block on that level represents the "gatekeeper" for those wishing to develop a new standard (New Work). It normally requires significant justification that the proposed subject should be standardized, that no other suitable standard (domestic, regional, or international) already exists and could be adopted, and that the human and other resources are available for the development of the standard. Moreover, it is most important that the SDO carefully scrutinize the *scope* of the proposed standard to make sure that it does not come into conflict with or duplicate existing standards or ongoing standard development efforts.

The two lower levels, the Technical Committees and the Subcommittees (sometimes called Working Groups) are not separate in all SDOs. It is at these levels that most of the actual work of developing a standard is performed, including the developing of consensus and compliance with the requirements of due process.

Development Process for Classical Methods (SDOs)

It is far better to adopt a proven standard than to develop a new one! Historically, the United States. has a poor track record of adopting International Standards. It was reported in 1989 that fewer than twenty International Standards were directly adopted in the United States. There may have been a larger number of roughly equivalent U.S. standards. U.S. representatives to ISO and IEC Technical Committees participated in the development of International standards but the United States failed to adopt them, much to the chagrin of the European Community (now European Union). This situation has rapidly improved since the late 1990s. U.S. Standards Bodies that have led in the adoption of International Standards include (among others):

American National Standards Institute (ANSI, via an Accredited SDO)

American Petroleum Institute (API)

American Society for Quality (ASQ)

American Society for Testing and Materials International (ASTMI)

Association for Advancement of Medical Instrumentation (AAMI)

Institute of Electrical and Electronics Engineers (IEEE)

InterNational Committee for Information Technology Standards (INCITS)

Instrumentation, Systems and Automation Society (ISA)

Telecommunications Industry Association (TIA)

Underwriters Laboratories (UL)

Some of the International standards adopted by the United States might be termed "boomerang standards," having been previously developed in the United States or elsewhere. It is worth noting that there are ISO/IEC guides (ISO/IEC 2005) for the regional or national adoption of International standards (or other standards) in addition to the guidance in Guide 2.

The development process is illustrated in the flow chart of Figure 4.2. The applicant proposing to develop a new standard provides a rationale supporting that proposal. This takes various forms in different SDOs, but in general a "business case" must be made to justify a new work effort. If the gatekeeper decides to accept the rationale for developing a new standard, the Subcommittee is given the go-ahead to proceed. If not, the effort is abandoned or possibly pursued in another standards forum.

How to get started right. The Working Group must perform some housekeeping operations such as the election of a chair and other officers, the creation of an outline of the proposed standard, and the scheduling of the work. Specific timeline goals must be set and adhered to if the completion of the development process is to be reasonably short. All the development plans for the effort should be widely publicized, and *users* as well as producers and others should be involved in a balanced way.

The makeup of the Working Group is also important. Members must be technically competent is the subject matter under development. Also, there must be leadership in the Working Group with people skills that keep the group focused and actively working to establish consensus. Sometimes (not often it is hoped), the state-of-the-art technical solution must give

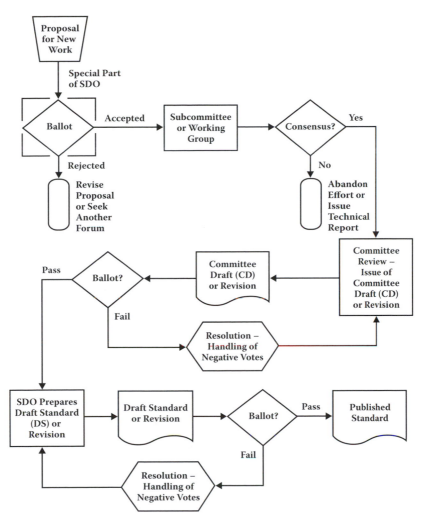

FIGURE 4.2
A typical SDO standard development process.

way to a political compromise! In some respects, the development of the Working Group process resembles a contract negotiation with give and take on all sides.

The Working Group must take full advantage of modern electronic tools to simplify and speed the development of a standard. ISO has developed a template that simplifies and speeds the development of ISO/IEC/ITU standards. SDOs and others hoping to have a standard become an International Standard will, presumably, use the ISO template from the outset.

One of the issues that is sometimes overlooked is the identification of the terminology to be used. Terminology is so important in standards that ASTMI requires, in a special Paragraph 3, that the terminology to be used in a newly developed standard be identified (by reference) or created and included when special terms are required. A guide to the selection of terms for use in standards development appears in Figure 4.3.

Once the Subcommittee has achieved a consensus on a draft, as evidenced by a ballot in some SDOs (including ANSI accredited SDOs), it is sent to the Committee for review and preparation of a Committee Draft (CD). A

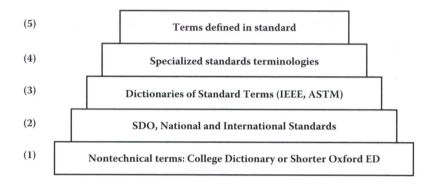

NOTES:

[1]   Do not use Shorter Oxford English Dictionary (ED) for Standards or Conformity Assessment terms!

[2]   ISO/IEC Guide 2, IS 17000, IEEE & ASTMI Terminology standards, etc. One of the more important of such standards is The International System of Units (SI): See, for example, NIST Special Publication 330 (2001), Barry N. Taylor, Ed.

[3]   Roles of dictionaries: insights into other related terms; help prevent duplication; help in forming specialized terms, help in information retrieval work. One of the more important of these is the IEC multilingual dictionary.

[4]   Sometimes terms in (4) are refined for special terms defined in standard; Terms in the International Electrotechnical Vocabulary (IEV) and ISO vocabularies must be consulted for possible use.

[5]   Define special terms in standards carefully: they should eventually migrate to (4), then to (3), etc.

FIGURE 4.3
Bottoms-up terminology guide for standards development, from the general (1) to the more specialized (5).

somewhat wider review and ballot of the CD then takes place. Pursuant to the due process concepts, attempts are made to resolve negative comments and votes. The modified (revised) CD is then reissued by the Committee for another ballot. This part of the process may take several circuits around the indicated loop before a Pass ballot is obtained. At some point, not often it is hoped, the CD may not achieve a consensus as required and have to be abandoned or issued as a document that is not a Standard.

The SDO then takes the successful Draft Standard and circulates it to an even wider audience for comments and voting. Sometimes the CD is balloted in parallel with the Draft Standard. It is rare that a Draft Standard fails to eventually issue as a published Standard. Again, if consensus is not achieved under the rules of the SDO, a Technical Report or other nonconsensus document may be issued.

In practice, the details of the Standards development process are more complex than indicated here. ASDs must adhere to the ANSI Essential Requirements. Due process requirements for American National Standards are found on the ANSI Web page (www.ansi.org) at the Library–Public Documents area.

Classical Standards Development Issues Covered Elsewhere

The federal government often gets involved in the development of a Standard by an SDO using the classical method. This is frequently a prelude to the use by the government of such a Standard in an Agency rule or Regulation. More on that aspect of Standards development appears in Chapter 11, "Federal Roles in Standards, Conformity Assessment, and Accreditation." There is also an amplification of the legal issues in Standards development, including the handling of Intellectual Property Rights (IPR, including patents) in Chapter 10, "Legal Issues in Standards and Conformity Assessment." Participation of the government in a standard's development is not, of course, required for the adoption of a Standard by an agency.

Consortia and Similar Methods of Standard Setting

A consortium is a group of companies or organizations that join (but do not merge) to achieve a goal that none of the individual companies can achieve on its own. Standards Setting Consortia (SSCs) are a recent development, mostly emerging in the 1980s. The motivation for the establishment of such consortia is the relatively long time it frequently takes for SDOs to develop a standard. Consortia are frequently able to set a standard in a relatively short time.

There is a perception that consortia work faster because they are not burdened with all the requirements of due process. Although this may be true in some cases, consortia generally observe most of the due process discipline of SDOs as well as some additional requirements as outlined by Updegrove (1995). Because of the wide variety of consortia methods in use, it is difficult to make general statements about them as was done earlier for SDOs. A standard setting consortium may not be as open as due process requires. The committee members are more of the same mindset than members of an SDO and so do not (typically) run into the naysayers that seem to populate SDOs. Also, of course, the standards setting consortia do not face the gatekeeper function of classical SDOs. This has advantages and disadvantages since the consortium's development work may duplicate other standards development efforts and may, in extreme cases, result in "turf wars" between competing interests. By focusing on *pre-competitive* standards setting, the consortium may avoid antitrust issues to some extent. If the interests of the larger members of a consortium overwhelm the needs and desires of the smaller firms, the resulting standard may be ignored or replaced by another standard that treats smaller firms more equitably.

One differentiator of standards developing consortia and similar fora is the ultimate aim of the standards developed. If the users of the standard have no motivation in having a consortium standard, a Publicly Available Standard or PAS, transposed to an International Standard, the development may be aimed differently than a standard that needs to be transposed to an International Standard to be effective. To illustrate, consider the Semiconductor Equipment and Materials International (SEMI)® standards, primarily "hardware" oriented, and the Open Group standards, primarily "software" standards and Certification.

SEMI is a trade association that, in addition to setting standards like a consortium, carries out a variety of educational and promotional activities. In the early days of semiconductor manufacturing, there were a variety of wafer sizes being developed by different manufacturers. This made the development and production of commonly used industry tools very expensive and wasteful. SEMI set wafer-size standards that brought order to a very disorderly situation (Brown 1995). This was a classic example of variety reduction! Since semiconductor manufacturing is a global enterprise, globally accepted standards were needed. SEMI expanded its standard setting activities to Europe and Japan in a set of "international standards" that covered chemicals, equipment, materials, and packaging among other subjects. The technology of flat panel displays also became a subject for SEMI standardization work. Because both producers and users were involved, SEMI was able to develop standards that readily met industry needs. Even though its standards are not aimed at higher-level recognition, SEMI hews to the ANSI due process model to minimize problems. Since companies are more directly involved, enhancing the possibility of dominance and anticompetitive activity, SEMI is very astute in avoiding even the appearance of collusion. SEMI's development process even includes a right to appeal (see SEMI Web page at www.semi.org).

The Open Group (OG) consortium includes many of the major companies in the field of information technology, with an emphasis on open standards for software. The Open Group is also involved in the Certification of software. The Open Group is impartial to vendors and technology. Its goal is to promote information flow across boundaries worldwide. To achieve its goals, the standards of the Open Group may be transposed to International Standards. Its Unix Specification (version 3, 2003) has been issued as ISO/IEC 9945:2003. The Open Group has also worked with IEEE, which has approved a UNIX specification as IEEE Std 1003.1-2001 (Open Group 2004). The Open Group Architecture Forum includes more than fifty members, including industrial firms, government agencies, and academic institutions. The standards of the Open Group are consensus driven in a sense with voting rights determined by membership category. The Open Group consortium operates under the legal umbrella of the U.S. National Cooperative Research and Production Act of 1993. More information on the Open Group can be found at www.opengroup.org. Probably the best known consortium is the World Wide Web Consortium (W3C) that develops the XML, HTTP, and other derivative standards used to operate the web. The URL for the World Wide Web Consortium is www.w3.org.

Internet Engineering Task Force (IETF) Methods

Some consider the IETF to be another consortium, but it doesn't really fit the consortium mold. The IETF differs from most consortia in its openness; anyone can participate (Alvestrand 2004). It is also unique in its mode of operation, doing most of its business on the Internet. The IETF has been described as a bottom-up process for building consensus for the Internet standards (referred to as Requests for Comments or RFCs) and network protocols. All efforts are aimed at making the Internet run as smoothly and reliably as possible. The emphasis is on speed as well as quality in the developed standards.

Another practice of the IETF that distinguishes it from most consortia is the requirement that at least two documented interoperable implementations of a specification must be submitted before it is given the status of a draft standard. The development is based on "rough consensus," but it has an appeals mechanism. The resulting standards are available free on the Internet, not sold as with many consortia. The IETF's operating income is derived from a variety of sources that contribute to the Internet Society (ISOC), the umbrella organization for the IETF. Like many consortia, the ISOC has grades of membership, depending on the extent of the financial report provided. A good but dated overview of the IETF process appears in (Crocker 1993).

Because of its outstanding success in a complex and fast moving environment, the IETF has shown that there is a new way to make highly technical standards very rapidly. It has also cooperated with the International
Telecommunications Union (ITU) to create a badly needed standard, the
Gateway Control Protocol (IEEE 1999). Some standards experts think that
the IETF is "lightyears ahead of any other standards groups" (Jacobs et al.
2001).

Conclusions

The appearance of consortia and other standards setting organizations has
shown that there are many options for setting standards, thus breaking
the mold of the "one-way" SDO methods of the past. New organizational
arrangements are showing up, including the IEEE Industry Standards
and Technology Organizations (IEEE-ISTO), which helps various kinds
of standards developers by using the standards expertise of the IEEE
(Kowalenko 2006). A good description of the IEEE-ISTO appears in Lefkin
(2000). Another standards developer, ASTMI, is breaking out of the con-
fines of traditional activities. ASTMI has recently developed, at the request
of the Federal Aviation Administration (FAA), basic standards covering the
deregulated light sport aircraft industry (Kern 2005). Some of the advantages and drawbacks of the three methods outlined earlier are discussed
in the following.

Classical (traditional) SDO methods are still the gold standard of standards development. Government can hardly be expected (in most cases) to
adopt standards set by other than full consensus methods incorporating due
process. It has been demonstrated that the SDO method can speed up the
development of a standard, given the right conditions, for example:

- Making the initial meetings long enough to produce a good outline,
  completely define terminology, and so forth
- Meeting often (at least monthly)
- Using development tools and "virtual meetings" to advance the work
- Sizing the standard appropriately (not too big, not too small)

In some cases the SDO method doesn't take that much longer than a consortium method. Some experts question the merit of other methods to save
time (Batik 1999). Batik showed that hastily constructed standards can be
very wasteful and expensive in the long run.

Consortia and similar methods are here to stay and clearly have advantages in some situations like the SEMI example described earlier and others. Consortia are frequently able to respond to market needs more quickly and in some cases have their standards transposed to International Standards.

The IETF way has some detractors (Talbot 2005). The "piece work" method of development, it is claimed, lacks the overall coordinating vision that was the hallmark of the Open Systems Interconnect (OSI) standard concept. The U.S. dominance of the Internet, it is reported, is prompting some countries to develop alternatives that might reduce global interconnections.

References

Alvestrand, Harald, 2004, The Role of the Standards Process in Shaping the Internet, *Proceedings of the IEEE* 92(9): 1371–1374.

Batik, Albert, 1999, What Price Speed? *Standards Engineering* 51(6): 1–5.

Brown, William, 1995, The SEMI International Standards Program … A Path to Global Markets, *Standards Engineering* 47(4): 1–5.

Crocker, D., 1993, Making Standards the IETF Way, *ACM StandardView* 1(1): 48–54.

Hohmann, Larry, 1989, Process Control in Standards Bodies, *IEEE Communications Magazine* 27(9): 59–61.

IEEE, 1999, Agreement Reached on Important New Standard for the Control of Internet Voice Gateways, *IEEE Communications Magazine* 37(10): 144.

ISO/IEC, 1994, Guide 59. *Code of good practice for standardization*, Geneva.

ISO/IEC, 2005, Guide 21-1:2005. *Regional or national adoption of International Standards and other International Deliverables*—Part 1: Adoption of International Standards; Guide 21-2:2005—Part 2: Adoption of International Deliverables other than International Standards, Geneva.

Jacobs, Kai, Rob Procter, and Robin Williams, 2001, The Making of Standards: Looking Inside the Work Groups, *IEEE Communications Magazine* 39(4): 102–107.

Kern, Tim, 2005., Case Study: Light Sport Aircraft Industry, *ASTM Standardization News* 33(12): 28–33.

Kowalenko, Kathy, 2006, Group Gives Helping Hand to Industry Standards Developers, *News from Around the IEEE*, February 8 (see www.ieee-isto.org).

Lefkin, Peter, 2000, IEEE-ISTO Offers a Complementary Forum for Standards-Related Activities, *IEEE Power Engineering Review* 20(1): 36–37.

Open Group, 2004, *The Open Group Platform*, The Open Group Conference, San Diego, CA, February 2–6.

Penenberg, Adam, 2003, *Tragic Indifference*, New York, HarperCollins, p. 206.

Schumny, Harald, 2000, The Standards Process, *Computer Standards and Interfaces* 22: 1–3.

Shanahan, Kevin, 2005, The Duties and Responsibilities of Subcommittee, Chairs, *ASTM Standardization News* 33(8): 32–33.

Sheahan, Carolyn, 2005, Managing the ASTM Balloting Process and Handling Negative Votes, *ASTM Standardization News* 33(8): 34–37.

Swankin, David, 1990, *How Due Process in the Development of Voluntary Standards Can Reduce the Risk of Anti-Trust Liability*, NIST/GCR 90/571.

Talbot, David, 2005, The Internet Is Broken, *Technology Review,* December/January, pp. 62–69.

Updegrove, Andrew, 1995, Standard Setting and Consortium Structures, *ACM StandardView* 3(4): 143–147.

5

Accreditation of Conformity Assessment Bodies (CABs)—ISO/IEC 17011

Introduction

Accreditation is closely related to Certification. Their definitions both start with "third-party attestation." A principal difference between Accreditation and Certification is that their objects are different: Conformity Assessment Bodies (CABs) for Accreditation, and "objects of conformity assessment" (e.g., products) for Certification. Another major difference between Accreditation and Certification is that Accreditation assessments focus on the management as well as the "technical" aspects of its objects.

The motivation for accreditation comes from two main sources. First, the mantra of international trade is "one standard, one test, accepted everywhere." In the past, foreign countries were hesitant to accept the results of local CABs, especially in critical areas such as safety. The accreditation of the local CABs gives foreign customers greater confidence in the test results of the CABs, especially given cooperation between domestic and foreign CABs and between domestic and foreign accreditors.

The second major motivation for accreditation of CABs comes from legal requirements, such as the following example of the National Electrical Code (NEC), a model code (standard) that has historically been adopted, in whole or in part, by most of the states in the United States, as well as counties and cities that have "electrical codes." The language "electrical testing laboratory that is nationally recognized" dates as far back, at least, as the 1978 edition of the NEC (NFPA 1978).

ISO/IEC 17011 was developed to replace three similar but different standards for *general* requirements for accreditation of CABs:

ISO/IEC Guide 58
ISO/IEC Guide 61
ISO/IEC/TR 17010

Example of an Accreditation System in the United States

A useful example of an accreditation system in the United States is shown in Figure 5.1 The Recognition (Accreditation) of Nationally Recognized Testing Laboratories (NRTLs) is performed by the Occupational Safety and Health Administration (OSHA) in the Department of Labor. NRTLs generally test for electrical safety to improve safety in the workplace (and elsewhere).

The underlined terms in Figure 5.1 are some of those defined in ISO/IEC 17011; the numbers in parentheses identify the terms and their definitions in that standard and are shown in the bottom part of Figure 5.1. These terms are superimposed on the NRTL model to illustrate the workings of the accreditation system. OSHA is the accreditation body and it performs assessments of CABs. The specific example of the CAB is Underwriters Laboratories (UL), with which most people are familiar. The assessment in this example is somewhat more legalistic, since a government agency is involved! Notices of Proposed Recognitions are published in the *Federal Register* so that OSHA may receive public input on the proposals. Absent a sustained opposition, and given a successful assessment pursuant to the regulations, which are basically similar to those of ISO/IEC 17011, OSHA recognizes a Testing Laboratory and designates it a Nationally Recognized Testing Laboratory. For completeness, the Conformity Assessment process by the CAB, detailed in Chapter 2, is also shown with its "output" of Certification.

The heart of this system is the assessment. Note in the definition that it is a *process*, like the development of a standard. It includes the application to the Accreditation Body (OSHA in this case) for accreditation and the onsite demonstration of the adequacy of its testing facilities, the competence of its people, and the suitability of its management.

Outline of the Requirements of ISO/IEC 17011

The general requirements of ISO/IEC 17011 are outlined as follows:

1. Scope: In addition to guiding accreditation bodies in their Assessment of CABs, the requirements of this standard also apply to peer evaluation processes for mutual recognition between accreditation bodies.
2. Normative references: Include ISO 9000-2000 (Quality management systems), ISO/IEC 17000 (Conformity assessment), and VIM:1993— International vocabulary of basic and general terms in metrology.
3. Terms and definitions (some of which appear in Figure 5.1).

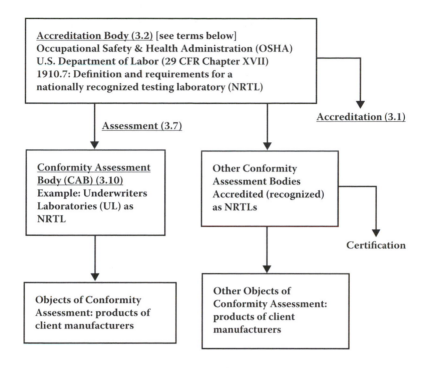

Accreditation Body (3.2) [see terms below]
Occupational Safety & Health Administration (OSHA)
U.S. Department of Labor (29 CFR Chapter XVII)
1910.7: Definition and requirements for a
nationally recognized testing laboratory (NRTL)

<u>Assessment (3.7)</u>

<u>Accreditation (3.1)</u>

Conformity Assessment
Body (CAB) (3.10)
Example: Underwriters
Laboratories (UL) as
NRTL

Other Conformity
Assessment Bodies
Accredited (recognized)
as NRTLs

Certification

Objects of Conformity
Assessment: products of
client manufacturers

Other Objects of
Conformity Assessment:
products of client
manufacturers

[Terms and definitions from ISO/IEC 17011]

3.1   accreditation — third-party attestation related to a conformity assessment body conveying formal demonstration of the competence to carry out specific conformity assessment tasks

3.2   accreditation body — authoritative body that performs accreditation

3.7   assessment — process undertaken by an accreditation body to assess the competence of a CAB, based on particular standard(s) and/or other normative documents and for a defined scope of accreditation

3.10   conformity assessment body — body that performs conformity assessment services and that can be the object of accreditation

FIGURE 5.1
A specific example of an accreditation system in the United States.

4. Accreditation body: Detailed requirements for accreditation bodies, including their legal status, structure, impartiality, confidentiality, liability and financing, and accreditation activity. Impartiality requires independence from both the CABs and their clients.

5. Management: Detailed requirements for management, including a management system, document control, correction of

nonconformities, preventive actions, internal audits, management reviews, and the handling of complaints.

6. Human resources: Requirements to assure suitable and adequate personnel, and personnel records.

7. Accreditation process: Covering the *process* of accreditation including criteria and information, handling of applications for accreditation, resource review, subcontracting of assessments, preparations for assessment, documents and records review, onsite assessment, analysis of findings, and assessment report. Decision making and granting of accreditation, appeals (due process again), reassessment and surveillance, extending of accreditation, suspending, withdrawing or reducing accreditation as well as records on CABs.

8. Responsibilities of the accreditation body and the CAB, including the obligations of the CAB, the obligations of the accreditation body, references to accreditation and use of symbols.

9. Bibliography: A partial menu of standards and guides to which a reference is made in the standard.

These requirements mirror, but go beyond, those in requirements for CABs, for example, ISO/IEC 17025 that contains requirements for testing and calibration laboratories.

Connections between Standards for Vocabulary, Conformity Assessment, Accreditation, and Other Applicable Standards and Documents

Many of the standards for conformity assessment show up in the bibliography of ISO/IEC 17011 as well. While several variations in arrangement are possible, Figure 5.2 shows how the following standards are coordinated:

(1)   ISO/IEC Guide 2—*Standardization and related activities*

(1.5) ISO/IEC Guide 59—*Code of good practice for standardization,* applicable to but not referenced in (1)

(2)   ISO/IEC 17000 is referenced in paragraph 12 of Guide 2.

(3)   ISO/IEC 17000—*Conformity assessment—Vocabulary and general principles*

(3.5) ISO/IEC Guide 60—*Conformity assessment—Code of good practice;* applicable to but not referenced in (3)

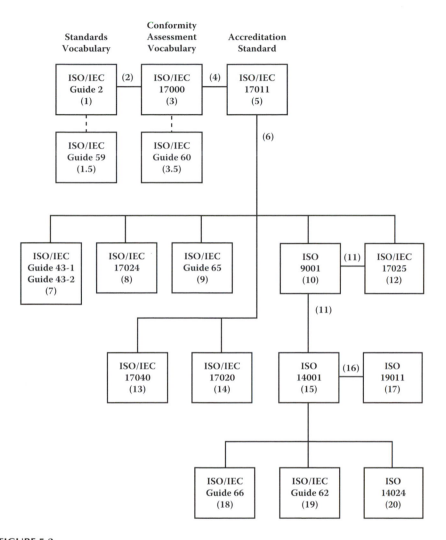

FIGURE 5.2
Connections between guides and standards for standards, conformity assessment, and accreditation.

(4)   ISO/IEC 17011:2004 is cited in the Bibliography of ISO/IEC 17000

(5)   ISO/IEC 17011, *Conformity assessment—General requirements for Accreditation bodies accrediting conformity assessment bodies*

(6)   Bibliography of ISO/IEC 17011

(7)   ISO/IEC Guide 43-1, *Proficiency testing by interlaboratory comparisons—Part 1: Development and operation of proficiency testing schemes;* ISO/IEC Guide 43-2, *Proficiency testing by interlaboratory*

comparisons—Selection and use of proficiency testing schemes by laboratory accreditation bodies

(8) ISO/IEC 17024:2003, *Conformity assessment—General requirements for bodies operating certification of persons*

(9) ISO/IEC Guide 65:1996, *General requirements for bodies operating product certification systems*

(10) ISO 9001:2000, *Quality management systems—Requirements*

(11) Coordination of requirements with ISO/IEC 17025 and ISO 1400

(12) ISO/IEC 17025:1999 (revised 2005), *General requirements for the competence of testing and calibration laboratories*

(13) ISO/IEC 17040:2005, *Conformity assessment—General requirements for peer assessment of conformity assessment bodies and accreditation bodies* (published after publication of ISO/IEC 17011)

(14) ISO/IEC 17020:1998, *General criteria for the operation of various types of bodies performing inspection*

(15) ISO 14001, *Environmental management systems—Specification with guidance* for use

(16) Citation of ISO 19011 in paragraph 5.7.1 of ISO/IEC 17011 and [4] of ISO/IEC 17000

(17) ISO 19011:2002, *Guidelines for quality and/or environmental management* systems auditing

(18) ISO/IEC Guide 66:1999, *General requirements for bodies operating assessment and certification/registration of environmental management systems (EMS)*

(19) ISO/IEC Guide 62:1996, *General requirements for bodies operating assessment and certification/registration of quality systems*

(20) ISO 14024:1999, *Environmental labels and declarations—Type 1 environmental labeling—Principles and procedures*

Summary

This is a relatively short chapter but more detailed coverage of actual accreditation systems is provided for the international and national areas in Chapters 6 and 8, respectively.

Reference

National Fire Protection Association (NFPA), 1978, National Electrical Code 1978, NFPA No. 70-1978.

6

International Standards, Conformity Assessment, and Accreditation

What is an International Standard?

To those of us who have bought into the longstanding use of the term International Standard and its definition in ISO/IEC Guide 2 (Chapter 1), this question seems strange indeed. As the connection between International Standards and trade has emerged, the General Agreement on Tariffs and Trade (GATT) urged the use of International Standards to avoid the hindrances to international trade called nontariff barriers of Technical Barriers to Trade (TBT). That policy became the Code of good practice for standardization (ISO/IEC 1994). The Code of good practice for standardization was set forth in ISO/IEC Guide 59, which became a seminal document in the development of *due process* in standards development, as described in Chapter 4 of this work. The Informative Annex A of Guide 59 included terms and definitions in relation to the GATT Agreement on Technical Barriers to Trade. The terms presented in the sixth edition of the ISO/IEC Guide 2:1991, *General terms and their definitions concerning standardization and related activities*, "shall, when used in this agreement, have the same meaning as given in the definitions of the said Guide." There was a modification of the ISO/IEC Guide 2 (1991) definition of an international standards organization.

The successor organization to the GATT was the World Trade Organization (WTO). When the issue of what is an international standard was recently presented to it, the WTO abandoned the Guide 2 definition and defined process requirements including the transparency, openness, impartiality and consensus, effectiveness and relevance, coherence, and the facilitation for developing countries to participate (Grove 2005). Thus, Grove makes the case that ASTMI standards can be called "international standards" by virtue of that compliance with the WTO process. Other authorities point to how widely used a standard is to determine if it can be called an "international standard" (Smoot 2002). Smoot asks, "Should international standards encompass only the documents from organizations like ISO or include any document found to have value?" Smoot further adds that "you can be sure that businesses that rely on standards prefer one good globally accepted standard." International standards have been called Global Standards (OTA 1992) and in Mackay (2000). Accordingly, the "international standard" might be called a Globally Accepted Standard (GAS) to minimize confusion.

The discussion about what constitutes an international standard is sometimes referred to as the "little i versus the big I" debate. The Europeans and others consider that only the International Standards produced by the ISO, IEC, and ITU are properly called International Standards, while the U.S. (consensus) position is like that articulated by Grove and others. The U.S. consensus position has been transformed into official policy by the U.S. Trade Representative (USTR), which has formally objected to the position that the WTO TBT Agreement identifies specific ISO, IEC, and ITU standards (USTR 2006). The classical ISO/IEC definition is espoused by many, including Willingmyre (1997). One problem with the redefinition of the little i "international standard" is that it cannot readily be determined if a given standard meets the dual requirements of the WTO process and wide (global) use. International Standards frequently have a wider consensus base than Globally Accepted Standards; that makes them attractive to regulators and to users of safety and similar standards where wider conformance tends to reduce a manufacturer's liability. Regulators apparently risk having regulations overturned if they ignore International Standards (Schonfeld and Dobinson 2004). Regional standards organizations like the European Union and the Pacific Area Standards Congress (PASC) consider the term international standard to mean ISO and IEC standards (Kushnier 2007). Ironically, an authority on European standards referred to ASTM standards as having de facto international status (Nicolas 1988).

A Modest Proposal

If the twin requirements of process and wide use are verified, as for example by a canvass of the ISO or IEC National Committees, then the standard should be "certified" as meeting the WTO requirements for a Globally Accepted Standard and submitted as a PAS document to the ISO or IEC. That would simplify things and ultimately preserve the integrity of the International Standards system. It would also be pedagogically sound and simplify standards terminology for newcomers. The alternative is a rather dismal prospect that all standards, however and wherever developed and used, would become "international standards," which may not be in the long-term interest of international trade. As before, I plan to use the initial capitals I and S to denote "real" International Standards and will reference the others with a lowercase i and s and quotation marks as is done earlier and for "certification" in Chapter 1. More on standards and trade appears in Chapter 7, "Regional Standards and Conformity Assessment."

Why International Standards?

If there were no international trade and no international communications, International Standards would hardly be necessary! But international trade and communications are imperatives, especially in modern times. Variety reduction applies to standards as well as to things. If a manufacturer had to build products conforming to various national standards, it would lose the economy of scale that results from variety reduction. It is a given that international trade drives International Standards. For at least fifteen years there has been a desire on industry's part to have "one standard, one test, accepted everywhere." The best opportunity for doing that is to use International Standards. The harmonization and maintenance of standards are expensive and time consuming. Those costs are minimized by the use of International Standards in Regional and National standards. The case for the use of International Standards is succinctly stated by the presidents of the IEC and ISO and the secretary-general of the ITU in Falk, Leong, and Tarjanne (1997) and by Kitzantides (1992) who addresses the question "Why International Standards?" The economic benefits or standards, as outlined in Chapter 3 and elaborated in Chapter 12, also apply of course.

What's in International Standards for us in the United States? Until recently, U.S. participation in International Standards setting and adoption was anemic. An Office of Technology Assessment (OTA 1992) report states on page 80:

> Failure to understand the implications of international standards can have serious consequences for U.S. industry. The U.S. machine tool industry is a case in point. For years the industry was able to thrive without regard to international standards. Industry practices became de facto standards because the U.S. market for machine tools was so large. In a global market, where there is intense foreign competition, this is no longer possible. Not being involved in the development of international standards or experienced in producing products to foreign specifications, the U.S. industry lost its competitive edge.

The use of International Standards can be a facilitator for a firm in international trade. A firm that adheres to International Standards can use them in:

- Research
- Design and development
- Marketing
- Purchasing
- Manufacturing
- Installation and service

International Standards have been accused of being barriers to trade (Cuschieri 1998). Lowell (1997) points out that American business leaders consider that local, national, or regional standards can be barriers to trade but not International Standards. As will be explored more thoroughly in Chapter 7, many regional standards are "adoptions" of International Standards so such regional standards should not represent trade barriers any more than International Standards do.

International Standards Need to be Globally Relevant

It has been stated that International Standards are not always globally relevant, that is, some such standards favor a certain region or are not applicable worldwide, thus serving as technical barriers to trade. The World Trade Organization addressed these and similar issues in the Committee on Technical Barriers to Trade document G/TBT/1/Rev.8. The document reinforced the principles outlined in ISO/IEC Guide 59 and elsewhere governing standards development. In Part IX, paragraph D, Effectiveness and Relevance, this document states:

> 10. In order to serve the interests of the WTO membership in facilitating international trade and preventing unnecessary trade barriers, international standards need to be relevant and to effectively respond to regulatory and market needs, as well as scientific and technological developments in various countries. They should not distort the global market, have adverse effects on fair competition, or stifle innovation and technological development. In addition, they should not give preference to the characteristics or requirements of specific countries or regions when different needs or interests exist in other countries or regions. Whenever possible, international standards should be performance based rather then based on design or descriptive characteristics.

ISO has recently addressed these "principles" as described by Steven P. Cornish (2005). Cornish amplifies the principles and gives some specific examples of their application.

Conformity Assessment in International Trade

Conformity Assessment can also be used as a trade barrier. To avoid that, the *Code of good practice for standardization* ensures that *standardized* Conformity

Assessment becomes a trade *enabler* if suitable standards and organizational arrangements are used (para. 3, Guide 59). Also important are Memorandums of Understanding (MoUs), Mutual Recognition Arrangements (MRAs), and appropriate Accreditation (Chapter 5). International Metrology standards, starting with the *International Vocabulary of Basic and General Terms in Metrology (VIM)*, and the Conformity Assessment Standards and Guides reviewed in Chapter 2 provide foundations in this area.

The GATT (WTO) Agreement on Technical Barriers to Trade (TBT) 1994 addresses conformity assessment issues as follows:

Article 5: Procedures for Assessment of Conformity by Central Government Bodies

Article 6: Recognition of Conformity Assessment by Central Government Bodies

Article 7: Procedures for Assessment of Conformity by Local Government Bodies

Article 8: Procedures for Assessment of Conformity by Non-Governmental Bodies

Article 9: International and Regional Systems

Conformity Assessment (ISO/IEC 17000 in Chapter 2) is a normative reference in ISO/IEC Guide 60:2004 *Conformity assessment—Code of good practice (second edition)*.

Guide 60 implements and supplements the WTO TBT Agreement and its subsequent modifications. Guide 60 is of special importance at the international level. To minimize technical barriers to trade at the international level, Guide 60 stresses *confidence building* through a variety of mechanisms. Confidence building is also stressed in a paper on transportability of conformity assessment (Mader and Haataja 1999). Guide 60 relies on ISO/IEC 17000 for terms and definitions. Guide 60 also reviews the role of normative documents and their requirements, the rules of the game for Conformity Assessment activities, the systems and schemes used in Conformity Assessment, and the handling of the results of Conformity Assessment. These rules are conceptually similar to the Guide 59 principles. For portability of Conformity Assessments, the "optional alternatives" in both the standards and the conformity assessment methodologies used must be somewhat limited.

A good example of an international MRA is the US–EU MRA in which the National Institute of Standards and Technology (NIST) has assumed accreditation oversight of U.S. Conformity Assessment Bodies that test for Electromagnetic Compatibility (EMC) and Radio and Telecommunications Terminal Equipment (R&TTE; NIST GSIG: Conformity Assessment Bodies Validated for the Operational Phase, http://ts.nist.gov/ts/htdocs/210/gsig/valcabs-1200.htm). Conformity Assessment Activities that are part of

International standards organizations will be described in connection with those organizations. A recent development of special interest is the September 2006 release of the new ISO/IEC 17021, *Conformity assessment—Requirements for bodies performing audit and certification of management systems.*

The International Accreditation Forum (IAF) has set a two-year deadline (September 2008) for accreditation bodies to comply with that standard (see Chapters 2 and 5).

The Principal International Standards Organizations: ISO, IEC, and ITU

There are several dozen International Standards Bodies (see Appendix 6), many of which are connected to the United Nations (UN), but the most important are the International Organization for Standardization (ISO), the International Electrotechnical Commission (IEC), and the International Telecommunications Union (ITU). The "principal" bodies are also called "apex bodies." The apex bodies reside in Geneva, Switzerland. The "non-treaty" International Standards Bodies are the ISO and the IEC. National and regional governments are not directly connected with the ISO and IEC. The ITU, on the other hand, is a "treaty" International Standards Body since it is a specialized agency of the UN. The practical effect of this situation is that official members of the ITU are nations and that the other members involved in the setting of telecommunications standards must be "endorsed" by the nation in which they are domiciled. In the United States, for example, working members of the ITU must be certified by the State Department (via ITAC; see Figure 6.1B) in order to participate in ITU standards setting.

The ITU is the oldest of the apex International Standards Bodies and dates its beginnings to the late 1800s following the invention of the telegraph. The IEC was formed in the early 1900s pursuant to an international meeting in St. Louis in 1904. The ISO was not formed until after World War II, although a legacy body, the International Federation of the National Standardizing Association (ISA), dates to 1926; it was lost in WWII. The scopes of the three bodies are:

- ISO—Most things that are not electrotechnical or telecommunications
- IEC—Most things that are electrotechnical in nature
- ITU—Telecommunications

Thus the areas of activity of these three organizations are complementary rather than competitive for the most part. There is ongoing close communication and cooperation of the apex bodies, which is facilitated by their common location.

Some other standards bodies are called "international" in some quarters but are not so identified here because they may not meet the ISO/IEC Guide 2 definition of an International standardizing body (4.3.2). Such bodies include the World Wide Web Consortium, the Internet Engineering Task Force, and the European Telecommunications Standards Institute (ETSI). ETSI probably comes closest to being an international standards body because of its large international membership and the worldwide use of its standards.

Overview of Standards, Conformity Assessment, and Accreditation Bodies

The principal International Standards, conformity assessment, and accreditation bodies are shown across the top row in Figure 6.1. The second row shows the Regional standards bodies and the bottom row depicts the National standards bodies. Only the major bodies and interactions are shown in Figure 6.1 to avoid the figure from becoming too busy. Figure 6.1A is Eurocentrist to a large extent since the most developed Regional standards bodies reside in Europe. Asian bodies, for example, are not shown except for the telecom bodies connected to the ITU and ETSI. The bodies identified only by abbreviations are identified more fully in the lists at the right-hand side of Figure 6.1B (also see Table 6.1). Common English names are usually used in lieu of the official national names. International Standards are also known as "norms," derived from the French. Maureen Breitenberg (1989) at NIST has produced a useful directory that gives information on international and regional standards developers and conformity assessment bodies.

Common Elements of the ISO and the IEC

Some authors refer to the "ISO/IEC System." There is in fact very close cooperation between the ISO and the IEC at all levels. The liaison between the two bodies is via the CEOs. The ISO/IEC Directives provide rules for the development and maintenance of International Standards:

- ISO/IEC Directives—Part 1. Procedures for the technical work
- ISO/IEC Directives—Part 2. Rules for the structure and drafting of International Standards

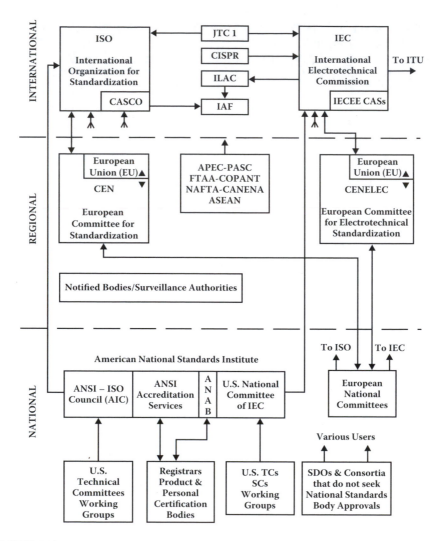

FIGURE 6.1A
Overview of standards, conformity assessment, and accreditation bodies.

- ISO/IEC Directives—ISO Supplement, Procedures specific to ISO
- ISO/IEC Directives—IEC Supplement, Procedures specific to IEC

The official languages of the ISO and IEC are English, French, and Russian. Many standards are published in English alone, and some in English and French (with Russian language to follow). Vocabularies, like ISO/IEC Guide 2 are published in English, French, and Russian, whereas the IEC Multilingual Dictionary defines terms in English, French, and Russian and includes terms in other languages.

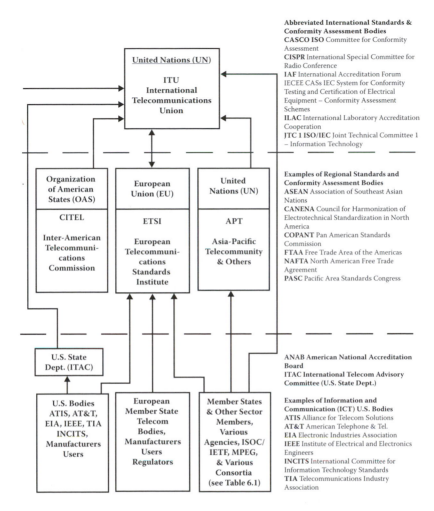

FIGURE 6.1B
Overview of standards and conformity assessment bodies.

ISONET and the International Classification for Standards (ICS)

Pursuant to the Uruguay Round Agreement at the WTO, it was decided to recommend that the ISO Standards Information System (ISONET) be the system of choice for communication between the ISO/IEC Information Center in Geneva and the other ISONET members to implement the Code of

TABLE 6.1

An Extension of Figure 6.1B: Examples of Consortia and Regional Bodies
Involved in ICT Standardization

Acronym	Organization
APT	Asia Pacific Telecommunity
ATM Forum	Asynchronous Transfer Mode (Consortium)
ATSC	Australian Telecommunication Standardization Committee
CEA	Consumer Electronics Association
DSVD	Digital Simultaneous Voice and Data Forum
ECMA	Formerly, European Computer Manufacturers Association
FSAN	Full Service Access Network
ISOC/IETF	Internet Society/Internet Engineering Task Force
ITSC	Interregional Telecommunications Standards Conference
MPEG	Moving Picture Expert Group (from ISO/IEC JTC 1)
OIF	Optical Internetworking Forum
PCCA	Portable Computer and Communications Association
TSACC	Canada, Telecommunications Standards Advisory Committee
TTA	Korea, Telecommunications Technical Association
TTC	Japan, Telecommunication Technology Committee

Good Practice for the Preparation, Adoption and Application of Standards in
Annex 3 to the Agreement on Technical Barriers to Trade.

In addition, an alpha(numeric) system for the language-independent classification system for identifying the subject matter of international standards is mandated. That is, the classification system used in the ISO and IEC
Catalogs of subject matter for the nine sectors into which most International
Standards fall:

1. General, including terminology, standardization, management,
   and science

2. Health, safety, and environment, including health care technology

3. Engineering technologies, including metrology and mechanics

4. Electronics, including information technology and
   telecommunications

5. Transportation and distribution of goods (including the container
   standard)

6. Agriculture and food technology

7. Materials technologies, including textiles, minerals, metallurgy, etc.

8. Construction, including materials and civil engineering

9. Special technologies including military, commercial equipment, and sports

The expanded numbering and decimalization of these basic sectors provides open standards classification systems in ISO and IEC of almost unlimited scope.

The Standards Development Stage Codes

Further, the WTO–ISO system called for a stage code system so that the stage of development of an ISO standard (and by extension all International Standards being developed) could readily be ascertained. The stage code would provide an alpha(numeric) description of the stage of development of a standard. Five stages of development were mandated as follows:

Stage 1:   The stage at which the decision to develop a standard has been made.

Stage 2:   The stage at which technical work has begun but for which the period of comments has not yet started.

Stage 3:   The stage at which the comment period has started but has not been completed.

Stage 4:   The stage at which the period for the submission of comments has been completed, but the standard has not yet been adopted. (Note that the use of "adopted" here is not the same as the ISO/IEC Guide 2 meaning. That term in ISO/IEC Guide 59, which implemented the WTO recommendations in Informative Annex B was changed to replace "adopted" with "formally approved.")

Stage 5:   The stage at which the standard has been "adopted."

An Example of the Project Stage Codes

Table 6.2 is an amalgamation of several such tables in the ISO/IEC Directives.

TABLE 6.2

ISO/IEC Project Stage Codes

Two Digit Project Stage	Name	Associated Document	Acronym
00	Preliminary stage	Preliminary work item	PWI
10	Proposal stage	New work item proposal	NP
20	Preparatory stage	Working draft(s)	WD
30	Committee stage	Committee draft(s)	CD
40	Enquiry stage	Enquiry draft, Draft International Standard	DIS
50	Approval stage	Final Draft International Standard	FDIS
60	Publication stage	International Standard, or Guide, Technical Report	ISO or IEC, or ISO/IEC
90	Review stage	Periodic review, IS to be revised, withdrawn, or confirmed	
95	Withdrawal stage	Withdrawl ballot; Notice of withdrawal	

Note: Many intermediate "substage" actions (not shown) are involved in progressing from each stage to the next.

The International Organization for Standardization (ISO)

The ISO, a nontreaty organization, is a federation of national standards bodies that includes most of the industrialized nations of the world. The acronym ISO, indicating equal as in isobar, promotes the concept of one vote for each national member body. As shown in Figure 6.2, there are three categories of members in ISO:

- Member bodies (with full voting rights)
- Correspondent members (without voting rights, for nations without mature standards bodies)
- Subscriber members (for small national economies with no national standards bodies)

At the end of 2007, there were 104 national member bodies, 43 correspondent members, and 10 subscriber members. There were 201 Technical Committees (TCs), 542 Subcommittees (SCs), and 2287 Working Groups (WGs).

The Council, an eighteen-member subset of the member body representatives, manages the operations of the ISO; it appoints the members of the Technical Management Board (TMB) and the chairpersons of the Policy Development Committees. The Council also settles on the budget of the Central Secretariat.

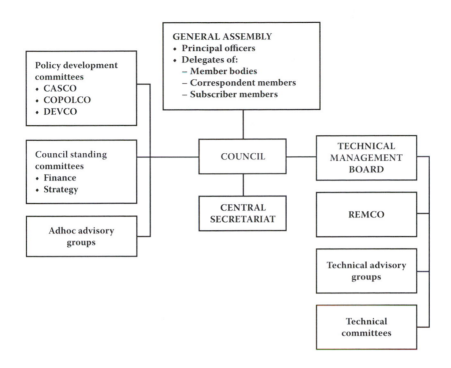

FIGURE 6.2
Organigram of ISO.

The Central Secretariat provides secretariat services (see Chapter 4) to the General Assembly, the Policy Development Committees, and the Committee on Reference Materials (REMCO). A reference material, it may be recalled, is a material or measurement system that defines one or more values of a quantity and serves as a reference, for example, a 100 ohm standard resistor (para. 6.1 of the VIM). Such reference materials are called Etalons in French. The TMB has twelve members taken from the member bodies.

The TMB advises the Council on the organization, coordination, planning, and programming the technical work of ISO. The TMB reviews new proposals for ISO technical activities and on the establishment and dissolution of Technical Committees. The TMB discharges ISO's responsibility for the ISO/IEC Directives. The TMB also monitors the work of all ISO Technical Committees and approves the scopes and programs of Technical Committees. The TMB controls the assignment of secretariats and subcommittees and appoints chairs of Technical Committees. The TMB resolves technical coordination issues between ISO and IEC Technical Committees. In addition, the TMB advises the secretary-general on cooperation with other international standardizing bodies. The TMB establishes registration authorities for the maintenance and implementation of international standards. The TMB establishes and dissolves Technical Advisory Groups (TAGs) and appoints their members and chairpersons. The TMB also establishes and dissolves committees on general standardizations and appoints their chairpersons. Strategic and technical advisory groups are established by the TMB as needed to oversee particular fields and explore standardization opportunities for ISO.

The Committee on Reference Materials (REMCO) is open to member bodies that are interested in its work and also to correspondent and subscriber body members as observers. REMCO establishes the definitions levels and classification of reference materials that are used by ISO. REMCO also formulates criteria to be applied for choices of reference materials, including legal aspects. REMCO proposes actions on reference materials required for ISO work and deals with matters on reference materials in relation to other international organizations and advises the TMB of actions to be taken. Other ISO and IEC committees such as CASCO (ISO Committee on Conformity Assessment), as well as International Standards Bodies including BIPM (International Bureau of Weights and Measures), ILAC (International Laboratory Accreditation Cooperation), and OIML (International Organization of Legal Metrology) liaise with REMCO.

Policy development committees are advisory committees established by the General Assembly with reporting to the Council. The ISO/CASCO is responsible for the Standards and Guides on Conformity Assessment that were covered in Chapter 2, especially IS 17000. CASCO studies the assessment of conformity of products, processes, and management systems and prepares appropriate standards and guides. It also reviews inspection bodies, certification bodies, and accreditation bodies. CASCO also promotes mutual recognition and acceptance of national and regional conformity assessment bodies and, as indicate in Figure 6.1A, oversees the International Accreditation Forum (IAF) and the International Laboratory Accreditation Cooperation (ILAC). CASCO also liaises with REMCO and DEVCO (ISO Committee on Developing Country Matters), among others. The ISO Committee on Consumer Policy (COPOLCO) promotes the means of helping consumers to benefit from standardization activities and of consumer participation in international standards development. COPOLCO advises the ISO Council on consumer views on standardization and conformity assessment work and to advise the Council on the policies for responding to consumer needs.

The Committee on Developing Country Matters (DEVCO) is another of the policy development committees. DEVCO's function is to identify standards and conformity assessment related needs of developing countries that generally cannot afford to pursue such activities on their own. DEVCO monitors the ISO Action Plan for developing countries and provides a forum for reviewing aspects of standards and related activities in developing and developed countries. DEVCO has Regional Liaison Officers stationed in various parts of the world, including Africa, the Caribbean, and Central America.

The last block in Figure 6.2 includes Ad hoc advisory groups, which are temporary groups set up to advise the Council as required.

ISO Standards

The ISO 9000 series (family) of standards on Quality Management and the ISO 14000 series of standards on Environmental Management have become famous because of their wide impacts. Some U.S. managers resisted the adoption of ISO 9000 standards on the theory that their company standards on quality management were superior to ISO 9000 (probably true in some cases). Some managements missed the fact that the variety reduction achieved by industry convergence on the ISO 9000 standards made registration and accreditation processes faster and better (and more economical) than would otherwise have been the case. The use of ISO 9000 standards by the suppliers of large manufacturers simplifies the evaluation of those suppliers. ISO 9000, Quality management systems—Fundamentals and vocabulary (ISO 9000: 2000) states in the introduction:

> The ISO 9000 family of standards listed below has been developed to assist organizations, of all types and sizes, to implement and operate effective quality management systems.
>
> - ISO 9000 describes fundamental of quality management systems and specifies the terminology for quality management systems.
> - ISO 9001 specifies requirements for a quality management system where an organization needs to demonstrate its ability to provide products that fulfill customer and applicable regulatory requirements and aims to enhance customer satisfaction.
> - ISO 9004 provides guidelines that consider both the effectiveness and efficiency of the quality management system. The aim of this standard is improvement of the performance of the organization and satisfaction of customers and other interested parties.
> - ISO 19011 provides guidance on auditing quality and environmental management systems.
>
> Together they form a coherent set of quality management system standards facilitating mutual understanding in national and international trade.

So much good material on ISO 9000 standards exists that it seems unnecessary to go into that subject deeper here. Suffice it to say that the American Society for Quality (ASQ) has been a leader in the U.S. adoption and application of ISO 9000 standards. A number of "derivative standards" for specific industry sectors have been developed, for example, in the automotive sector. The ISO 14000 series of standards are far too numerous to be surveyed here. The European Union is driving regional and international environmental requirements with its WEEE (Waste Electrical and Electronic Equipment) and the RoHS (Restrictions on Hazardous Substances) Directives, which will be discussed in Chapter 7. An excellent introduction to ISO 9000 and ISO 14000 standards appears in the *International Standards Desk Reference*

(Zuckerman 1997). The ISO 14000 standards started out as another redundant layer on the ISO 9000 series, but the basic parts of the two series were coordinated early on.

ISO Standards Development Processes

The standards development processes in ISO are elaborations of the SDO (Standards Developing Organization) process shown in Figure 4.2 of Chapter 4. The development of ISO standards has been made easier and faster by the recent introduction of the "ISO Templates." The details of the process are too articulated to go into here, but the stages identified in Table 6.2 provide some insight into the process when it is realized that each stage (from 00 to 95) is implemented by as many as a half-dozen substages. As shown in Figure 6.1, the U.S. Technical Advisory Groups (TAGs) and Working Groups (WGs) interface directly with their counterparts from other national members of ISO. Individual participants have to be "certified" by ANSI to the ISO activity involved and must be suitably supported by their employers. These individuals are generally recognized by their peers as experts in the subject matter of concern. More information on ISO can be found at its web site: www.iso.ch.

ISO/IEC Joint Technical Committee 1 (JTC 1)

ISO/IEC Joint Technical Committee 1 (JTC 1) combined several previously overlapping Technical Committees in information technology in ISO and IEC. JTC 1 has more than 2000 standards in information technology under its wing and cooperates with numerous other Technical Committees and Subcommittees in both ISO and IEC. JTC 1 also supports the development of telecom standards by working with the ETSI (European Telecommunications Standards Institute) and the ITU (International Telecommunications Union). The Technical Committees of JTC 1 cover, among other things, coded character sets, software and programming, image processing, interconnection of information technology equipments, office equipments, coding of audio and pictures, and user interfaces. The Moving Pictures Expert Group (MPEG) is a Working Group of JTC 1.

A historical focus for ISO was on developing the standards connected with the Open Systems Interconnect (OSI), including ISO 7498. The OSI standards development was too slow and it was overtaken by the Internet Engineering Task Force (IETF) standards.

John Voelcker (1986) has provided a good tutorial on the OSI. Ken Krechmer (1996) points out that fixed, nonoverlapping Internet standards are more important at the lower levels, layers 3 and 4, where protocols support end-to-end communication.

JTC 1 is currently developing biometric (face recognition, etc.) standards for security systems, local area network (LAN) standards, and programming languages like C#. The IEEE 802 standards have been fast-tracked into the

ISO/IEC system for nearly twenty years. The ability of JTC 1 to transpose consortium standards, like the Unix standard mentioned in Chapter 4, into ISO/IEC standards give it a key role to play. More on JTC 1 appears at www. jtc1tag.org and in an IEEE Computer Society publication (Moore 1998).

The IAF/ILAC Partnership

As shown in Figure 6.1A and mentioned previously, the International Accreditation Forum (IAF) and International Laboratory Accreditation Cooperation (ILAC) operate with the "supervision" of ISO/CASCO. The IAF requires Accreditation Bodies (Chapter 5) to operate at high technical and ethical levels, thus assuring users of those bodies, such as ISO 9001 Registrars, that accreditations will be respected worldwide. The IAF achieves that goal by the IAF Multilateral Recognition Agreement (MRA). The IAF registrars are mostly focused on the registration/accreditation of Conformity Assessment Bodies involved in ISO 9001 and ISO 14001 activities. The IAF uses International Standards in this process, especially ISO/IEC 17021:2006. IAF has given Certification bodies two years to make the transition to the new ISO/IEC 17021.The URL for the IAF is www.iaf.org.

The International Laboratory Accreditation Cooperation (originally Conference) deals with testing and calibration laboratories as well as Inspection laboratories. ILAC also has a network of members that agreed to an "Arrangement" that took effect in 2001. The members of the agreement operate peer review processes to assure their continuing competence. The aim is the motto "product tested once and accepted everywhere." ILAC also uses appropriate International Standards, especially ISO/IEC 17025:2005, *General requirements for the competence of testing and calibration laboratories.*

The two organizations, IAF and ILAC, have formal agreements with ISO and also have a joint licensing agreement for their approval marks. The URL for ILAC is www.ilac.org. The American National Accreditation Body (ANAB) set up by ANSI and the American Society for Quality (ASQ) is a member body of the IAF.

The International Electrotechnical Commission (IEC)

The IEC, a nontreaty organization, has its roots in the United States. In 1904, twenty-five famous electrical engineers and scientists from around the world gathered at the St. Louis (U.S.) International Electrical Congress and concluded that the technical societies of the world should develop an international standards group (Mason 1980). The Mason paper shows a group picture of the delegates, which included such famous Americans as A. E. Kennelly, S. W. Stratton, and Elihu Thompson. The IEC was actually formed two years later in 1906 in London; its first president was British scientist Lord Kelvin. After World War II, the IEC moved from London to Geneva in 1948, where the ISO and other international organizations were being founded. In 2006, the IEC celebrated its hundredth birthday, "the electric century."

As shown in Figure 6.3, the Council, consisting of the National Committees, controls the operation of the IEC. The Council is the policymaking organ of the IEC and sets long-term objectives. The Council delegates the management of IEC work to the Council Board. The Council Board, in turn, controls the Standardization Management Board and the Conformity Assessment Board. The Council Board includes the IEC officers and the voting members of the Council. The Central Office (the Executive) is lodged in the Council as indicated. The Council Board may establish Management Advisory Committees, as indicated, but these are like the "Detailed Administrative Functions" shown in Figure 4.1 and are not of particular interest here.

The Standardization Management Board, as indicated in Figure 6.3, manages the IEC consensus standards work. It is responsible for the establishment of the Technical Committees and Subcommittees as well as the determination of Technical Committee scopes of standardization. The Standardization Management Board is also responsible for IEC coordination with other international standards organizations and the production schedules for Technical Committees and Subcommittees. The Standardization Management Board has

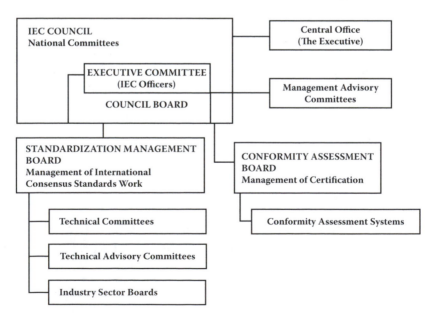

FIGURE 6.3
Organization of the IEC.

a Technical Advisory Committees on such subjects as safety, electromagnetic compatibility, and the environment. Industry Sector Boards include executives with market expertise in the areas of electricity transmission and distribution, industrial automation systems, and the infrastructure of telecommunications.

The Conformity Assessment Board provides the management of the IEC conformity assessment activities. The Conformity Assessment Board members are elected by the Council and include one member from each of the three Conformity Assessment Schemes. Although the details of their operation differ, the overall goal of the IEC Conformity Assessment Schemes is to encourage world trade by reducing nontariff barriers by providing one test, one certification, and worldwide acceptability:

- IECEE—IEC System for Conformity Testing and Certification of Electrical Equipment (the CB Scheme) and the CB–FCS Scheme
- IECEx Scheme—IEC Scheme for Certification to Standards for Electrical Equipment for Explosive Atmospheres
- IECQ Scheme—Provides conformity assessment and certification for electrical/electronic components

The IEC also certifies Photovoltaics (PV GAP) on a Global Approval Program.

Note: The term "scheme" has a negative connotation in the United States so the word is little used here in the same sense as used in the IEC schemes. Since the IEC Conformity Assessment Schemes have no counterpart in the ISO, they will be described in a little more detail. The following information on the IECEE Scheme is based on the CB Bulletin No. 108.

The IECEE CB Scheme

The name IECEE is short for the IEC System for Conformity Testing and Certification of Electrical Equipment. The official name of the CB Scheme is Scheme of the IECEE for Mutual Recognition of Test Certificates for Electrical Equipment. The rules of the CB Scheme are published in three IEC standards:

- IECEE 01—Basic Rules and Procedures of the IECEE
- IECEE 02—Rules and Procedures for the CB Scheme
- IECEE 03—Rules and Procedures for the Full Certification Scheme (FCS)

The CB Scheme is founded on the applicable International (IEC) Standards for the safety of electrical and electronic equipment. The scheme is a multilateral agreement between certification bodies in participating IEC member countries whose national product safety standards are "adoptions" of the appropriate IEC safety standards or are sufficiently well harmonized with those IEC standards. Where there are national differences between the national standard and the IEC standard, the differences are declared as part of the CB Test Certificates generated by the National Certification Bodies (NCBs). The CB Test Certificates are based on, and must be accompanied by, *CB Test Reports*. CB Test Reports are, in turn, based on type tests, which document compliance of a product with each requirement of the applicable IEC standard. The type testing may be conducted at the NCB laboratory or at the manufacturer's testing laboratory under several different arrangements. The CB Test Certificates and Reports are then used by the manufacturer to obtain product safety approval (Certification) marks in the countries participating in the CB Scheme. The procedural steps are too detailed to be covered here, but the use of the CB Scheme avoids the multiple testing and Certification that were historically required to obtain safety approval marks in the several countries of interest. The CB Scheme serves importers as well as exporters in the support of free trade. U.S. participation in the CB Scheme was based on UL 1950, a standard that was sufficiently well harmonized with the (then) IEC 950 (now IEC 60 950-1, etc.).

The IECEE CB–FCS Scheme is a later extension of the basic CB scheme. The CB–FCS Scheme is based on Conformity Assessment Certificates as the means of acquiring national product safety Certification marks. The CB–FCS Scheme is a product certification system pursuant to ISO/IEC Guide 67—*Conformity assessment—Fundamentals of product certification*. This system focuses on the quality management system of the manufacturer and relies on product sampling of the affected product in the marketplace or at the manufacturer, and is characterized by ongoing surveillance of the Certified product as well as the manufacturer's ISO 9000-based quality system. The Follow-up program is designed to assure the suitability of the design and construction of the Certified product. By focusing on the declared differences between National Certification Bodies, the CB-FCS Scheme harmonizes safety approvals and reduces costs for manufacturers to obtain Certification in other countries. Like the CB Scheme, it is based on a multilateral agreements (MLAs).

To qualify for the CB–FCS Scheme, manufacturers must have a track record as CB scheme participants and comply with ISO/IEC Guide 65, ISO/IEC 17025, ISO/IEC 1720, ISO/IEC 17030, and ISO/IEC Guide 27.

IECQ–IEC Quality Assessment Systems for Electronic Components

The goal of the IECQ is to promote international trade in electronic components by certifying the quality of such components to International (IEC) Standards, specifications, and other documents. IECQ also Certifies processes and related materials. Typical components that are Certified under the system include:

- Integrated circuits

- Electrooptic components (optical isolators, etc.)
- Printed wiring boards (PWBs)
- Electromagnetic and electromechanical components
- Passive components (resistors, capacitors, etc.)
- Wires and cables

The need for components of measured quality and reliability arises in a number of situations, for example, when the component is a critical safety component, such as an optical isolator, which is relied on for electrical safety. Other applications of the system include avionic and military systems where reliability is crucial.

We already had component certification systems in the United States including UL Recognized Components and MIL SPEC QPLs (Qualified Parts Lists) but these certifications were not widely recognized in world trade. To say that the IECQ Scheme was met with little enthusiasm in the United States is understatement!

The IECQ system utilizes ISO 9001 but goes beyond that to assure that quality and reliability goals are met. The Member Bodies of the IECQ system are the National Authorized Institutions (NAIs), which provides country-wide management, and the Supervising Inspectorates (SIs), which in turn works with the U.S. Electronic Components Certification Board (ECCB). UL is a Supervising Inspectorate along with several other certifiers.

One of the major areas of activity for the IECQ now is the Certification that components are lead free (and free of other heavy metals and hazardous substances). This is in response to California legislation as well as Directives of the European Union:

- European Union Restriction on Hazardous Substances (RoHS)
- European Waste Electrical and Electronic Equipment (WEEE)

The standard for coping with these issues was produced by the Electronic Industries Association (EIA) working with the Electronic Component Certification Board (ECCB). That standard was adopted by the IECQ, which is also developing additional standards to address the environmental problems. The IEC Technical Committee is also involved in addressing the environmental issues in electronics products. The ASTMI has recently developed a standard in this area and the UL Recognized Component system also Certifies that components are lead free. A component Certification of any kind can, of course, be a useful tool for configuration management.

The IECEx Scheme

Explosive atmospheres have long presented safety problems at, for example, petroleum and petroleum distillates handling in many areas (refueling), hospital operating rooms, grain handling and storage facilities, underground coal mines, and so forth. These are called Ex areas, in IECEx parlance. Equipment that is used in those (hazardous location) areas is called Ex equipment. The main goal of the IECEx Scheme is to provide International testing and Certification for Ex equipments that can be safely used in Ex areas. Certified Ex equipment is listed in a database for easy identification by potential users.

The IECEx Rules for the operation of the scheme are:

- IECEx 01 IEC Scheme for Certification to Standards relating to Equipment for use in Explosive Atmospheres (IECEx Scheme)— Basic Rules
- IECEx 02 IEC Scheme for Certification to Standards relating to Equipment for use in Explosive Atmospheres (IECEx Scheme)
  - IECEx 02—Equipment Certification Program covering equipment for use in explosive atmospheres—Rules of Procedure
  - IECEx 03—IEC Scheme for Certification to Standards relating to Equipment for Use in Explosive Atmospheres (IECEx Scheme)
  - IECEx Certified Service Facilities Program covering repair and overhaul of Ex equipment—Rules of Procedure

There are a number of IEC standards involved, many from TC 31. The principal standard is the family of standards IEC 60079, IEC 61241, and IEC 61779. It is worth noting that UL has adopted most of the 60079 series in its own standards of the same numbers. As in the IECEE Scheme, national differences must be declared for testing and Certification. Those Certified must be able to test and assess samples. Manufacturing facilities are assessed and audited on a continuing basis. Because Ex equipment tends to be expensive, it is frequently repaired and overhauled rather than being replaced. To make sure that repairs do not impair the safety of the original equipment, part of the IECEx process involves Certification of Service Facilities. The IECEx Certificate and Mark thus provides a transportable approval valid in all participating countries.

CISPR (International Special Committee on Radio Interference)

The rapid expansion of radio in the early 1930s resulted in an increasing problem of interfering signals for radio receivers. A group of experts and interested international organizations was established to address the issue. That organization was named CISPR when it became part of the IEC several years later. The term "special committee" was used because the makeup of the group did not fit the "standard" IEC Technical Committee (see CISPR 10). In addition to the usual IEC National Committee members, the membership of CISPR includes (among others):

- CIGRE   International Conference of Large Electric Systems
- EBU   European Broadcasting Union
- ETSI   European Telecommunications Standards Institute
- EUROELECTRIC   Union of the Electricity Industry

Liaison arrangements exist (among others) between CISPR and:

- IEC/TC 69   Electric road vehicles and electric industrial trucks
- ICAO   International Civil Aviation Organization
- ITU   International Telecommunications Union
              (CISPR is actually an international organizational
              member)

CISPR is composed of several subcommittees as follows:

- SC CIS/A   Radio interference measurements and statistical
             methods
- SC CIS/B   Interference relating to industrial, scientific, and medi-
             cal radio-frequency apparatus; other heavy industrial
             equipment; overhead power lines; high voltage equip-
             ment; and electric traction
- SC CIS/D   Electromagnetic disturbances related to electric/elec-
             tronic equipment vehicles and internal combustion
             engine powered devices
- SC CIS/F   Interference relating to household appliances, tools,
             lighting equipment and similar apparatus
- SC CIS/H   Limits for the protection of radio services
- SC CIS/I   Electromagnetic compatibility of information technol-
             ogy equipment, multimedia equipment, and receivers

CISPR documents include Basic Standards, Generic Standards, and Product Standards. Basic Standards are produced by Subcommittee A of CISPR and by IEC TC 77, and focus on standards for measurement methods but do not include prescribed emissions or immunity limits. Basic Standards are applicable to almost all of the CISPR documents.

Generic Standards are based on particular environments, for example, residential and commercial. Generic standards include minimum requirements (essential requirements) in areas where specific standards do not exist. The code word "essential requirements" usually refers to an EU Directive (more on that in Chapter 7). Generic standards include some IEC TC 77 standards in the IEC 61000 series.

CISPR Product Standards address emissions and immunity limits as well as the methods of measurement used to address a range of products. Prominent among these is CISPR 22 covering emissions of Information Technology Equipment and CISPR 24 for the immunity requirements of ITE.

IEC–EMC Standards

IEC TC 77 (Electromagnetic compatibility) has three major subcommittees:

- SC 77A—Low frequency phenomena
- SC 77B—High frequency phenomena
- SC 77C—Immunity to high-altitude nuclear electromagnetic pulse (HEMP)

The standards produced by these subcommittees are intended to protect the electrical mains. One of the prominent, and more controversial of these standards, is IEC 61000-3-2: Limits for harmonic current emissions. This standard is of special concern to equipment that employ switch mode power supplies (SMPSs) as they tend to generate substantial power frequency (odd) harmonics that adversely affect the ac (alternating current) mains.

EMC standards have become more diffused with the increasing complexity of technology. For example, the EMC requirements for medical devices in IEC 60601-1-2 include electromagnetic requirements and tests.

The International Telecommunications Union (ITU)

Unlike the ISO and IEC, the ITU is a treaty organization, which means that U.S. participants in its activities must go through the U.S. Department of State (see Figure 6.1B). The ITU became an agency of the United Nations (UN) in October 1947 to assist in the management of the increasingly complex issues of international wired and wireless communications. The ITU was transferred from Bern, Switzerland, to Geneva that year. The ITU was reorganized into its present form at the 1992 Plenipotentiary Conference in Nice, France. The ITU has six official languages: Chinese, English, French, Russian, Spanish, and Arabic. Spanish, French, and English are the official languages. The rapid expansion of all kinds of communication in the last few decades—including fiber optic cable, satellite communications, cellular radio, the Internet, and the digitization of all kinds of communication—and the expanded need for regulatory control of the electromagnetic spectrum has made the ITU the focal point for ensuring end-to-end communication in telecommunications systems and spectrum management for the increasing use of radio frequencies worldwide. A rather detailed history of the ITU up to the early 1980s appears in a compendium titled *Spectrum Management and Engineering* (Matos 1985).

Because of the merger of "information technology" and communications, the term recently applied to this area is information and communication technology (ICT). The ITU has a distributed decision-making structure that relies on *Conferences* at international and regional levels. The ITU Study Groups are analogs of the Working Groups in ISO/IEC environments. As of early 2007, there were 191 national members of the ITU, 515 sector members (including all 3 sectors), 79 regional and other members, 11 regional telecom members, and 5 Intergovernmental Organizations Operating Satellite Systems. The modus operandi of the ITU is for a study question to be formulated at a conference, or elsewhere, and given to an appropriate *study group* for the development of a Recommendation (a standard). There are two basic processes available:

- Traditional Approval Procedures (TAP) with a nine-month approval cycle following the completion of the Study Group Work
- Alternative Approval Procedure (AAP) with a four-week (minimum) approval cycle following the completion of the work by the Study Group

The ultimate authority at the ITU is the Plenipotentiary Conference that includes delegates from Member States. The Plenipotentiary Conference normally meets every four years but can meet more often if necessary. The Plenipotentiary Conference develops policies, organizational aspects, and activities of the ITU. The Plenipotentiary Conference provides the broad management functions of the ITU including (among other things) the setting of dues, and the election of Council members, the Secretary General, Deputy, and Sector Directors. During the (usually) four years between Plenipotentiary Conferences, the Council provides governance for the ITU. The other parts of the organization include the Secretary General, the General Secretariat, the World Conference on International Communications, and the three sectors identified as ITU-D, ITU-T, and ITU-R.

Council: The Council is a subset of the Member States and members are elected so as to provide regional representation of the five world regions of the ITU. The Council is responsible for plans and policies as well as the daily operation of the ITU. It approves budgets and programs and sees to it that the overall goals of the union are achieved in accordance with the Union Constitution and the decisions of other Conferences, as required.

General Secretariat and Secretary General: The General Secretariat manages the financial aspects of the ITU activities and provides conference services, information services, and planning services. It also supplies legal advice and personnel services. Managing the many meetings and conferences is a major function of the General Secretariat. The General Secretariat provides documentation and translations, and publication support for the ITU. The General Secretariat prepares the draft budget for the union. The General Secretariat supports the Secretary General, the legal representative of the ITU, and custodian of ITU treaties, and provides advice on staff development and related matters.

World Conference on International Communications: The Plenipotentiary Conference sets World Conferences to review and revise, if necessary, the International Telecommunications Regulations. The regulations are treaty arrangements covering the operation of Public Communications Services as well as the supporting infrastructure. The International Telecommunications Regulations provide the focal point for telecom administrators (government entities) and telecom system operators. They also cover the interconnection and interoperability of the international ICT systems.

The three sectors—ITU-D, ITU-T, and ITU-R—are explained in more detail in the following sections.

The Telecommunication Development Sector (ITU-D)

It was recognized early that the availability of suitable telecommunication infrastructure was a key element in the economic and social development of less developed nations. The Organization for Economic Cooperation and Development (OECD) was a key player in that area. The UN declared 1983 as "World Communications Year: Development of Communications Infrastructure" and named the ITU the lead agency for international coordination of those issues. The Telecommunications Development Bureau (BDT) provides the executive function of ITU-D, and the BDT Director participates in telecommunications development conferences and other activities aimed at providing "appropriate technology" for communication to developing countries. As an example of an appropriate technology, studies have been performed on the potential of communication satellites for providing communications to rural and remote areas. The BDT Director works with the other ITU sectors and the other organizations involved in that effort. In addi-

THE INTERNATIONAL TELECOMMUNICATION UNION (ITU)

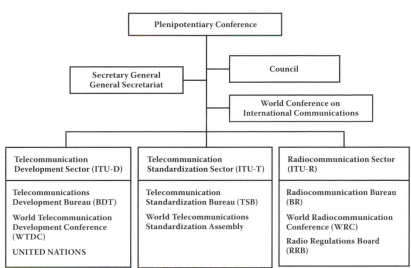

FIGURE 6.4
Simplified organization of the International Telecommunications Union (ITU).

tion to the OECD, the ITU-D works with the United Nations Development Program (UNDP) and other UN agencies.

The World Technical Development Conference (WTDC) meets every four years to review the progress made by developing countries in the establishment and improvement of ICTs. These reviews support the action plans for improving ICT infrastructure in developing countries. Regional Preparatory Meetings (RPMs) in the several ITU regions of the world are held prior to the WDTC to plan for development of ICT on a regional basis as well as a national one.

The Study Groups of the ITU-D are:

Study Group 1: Telecommunication development strategies and policies

Study Group 2: Development and management of telecommunication services and networks and ICT applications

The ITU Telecommunication Standardization Sector (ITU-T)

The task of the ITU-T (Figure 6.4) is to develop standards (Recommendations) on a variety of issues in worldwide communications of all types. ITU-T succeeded the CCITT (International Telegraph & Telephone Consultative

Committee). The Telecommunication Standardization Bureau (TSB) provides the executive function of the sector. It is headed by a Director elected by the Council. The Director participates in the activities of the World Communication Standardization Assembly and generally oversees the ITU-T Study Groups and organizes the attendant meetings and conferences. The ITU-T Study Groups include the following:

Study Group 2:   Operational aspects of service provision, networks, and performance

Study Group 3:   Tariff and accounting principles including related telecommunication economic and policy issues

Study Group 4:   Telecommunication management

Study Group 5:   Protection against electromagnetic environment effects

Study Group 6:   Outside plant and related indoor installations

Study Group 9:   Integrated broadband cable networks, and television and sound transmission

Study Group 11: Signaling requirements and protocols

Study Group 12: Performance and quality of service

Study Group 13: Next Generation Networks (NGN)

Study Group 15: Optical and other transport network infrastructures (e.g., IEEE 802)

Study Group 16: Multimedia terminals, systems, and applications

Study Group 17: Security, languages, and telecommunication software

Study Group 19: Mobile telecommunication networks

The ITU-T Study Groups are governed by the Telecommunication Standardization Advisory Group (TSAG). It is quite remarkable how ITU-T standardization has achieved so much so quickly with the assistance of ISO, IEC, IEEE, ETSI (European Telecommunication Standards Institute), IETF, JTC-1, MPEG, and other national, regional, and international standards developers. Some impediments such as IPR (intellectual property rights) have slowed the work at ITU; IPR issues are discussed at length in Chapter 10.

The ITU Radiocommunication Sector (ITU-R)

The rapid development of radio in the late 1920s and early 1930s resulted in increasingly severe problems of radio frequency interference (RFI). Until the ITU was reorganized along present lines, the RFI issues were handled by the CCIR (International Radio Consultative Committee). International arrangements addressing the RFI issue were the province of the International

Frequency Registration Board (IFRB). The 1912 Table of Frequency Allocations, circa 1912, was declared mandatory per treaty at the 1947 ITU Conference in Atlantic City.

As shown in Figure 6.4, the ITU-R includes, among other things, a Radio Communication Bureau (BR), a World Radiocommunication Conference (WRC), and the Radio Regulations Board (RRB), replacing the IFRB. The mission of the ITU-R is to provide RF spectrum management in an era of competing needs for spectrum space by mobile telecom (including cell phones), wireless Internet access, communication satellites, wireless internet access, radio astronomy, and a host of other applications.

The *Radio Communication Bureau* provides the executive function via an elected Director. The Director is responsible for the operation of the ITU-R and also serves on the Radio Regulation Board. Most of the issues that arise in the operation of the Radio Regulation Board can be resolved by the application of Radio Regulations. An appeals process is available where such issues cannot otherwise be resolved. The Director also provides support for the regional and world radiocommunication conferences that address spectrum management issues. Those issues include the efficient use of a limited (spectrum) resource as well as the avoidance of harmful interference, which degrades a radio service. The bureau also provides technical information to members and to the Telecommunication Development Bureau for assisting developing countries. The Radio Communication Bureau also maintains the Master International Frequency Register. The bureau has a staff of engineers and other specialists who assist the Director and also supports the various assemblies, study groups, and conferences.

The *World Radiocommunication Conferences*, also known as the World Administrative Radio Conferences (WARCs), are one of the key methods for developing solutions to radio problems. One of the such conferences, held in 1979, is known as WARC-79. The WARC-79 Conference addressed the new spectrum management issues involving satellite communication and other issues. A good summary of WARC-79 appears in Kirby (1980). Follow-up reports on the 1992 WARC appear in several papers in *IEEE Spectrum* for February 1992 (which also includes a historical sketch of the ITU).

The *Radio Regulations Board* (RRB) represents the ITU's five regions: Americas, Western Europe, Eastern Europe, Africa, and Asia and Australasia. RRB members serve independently and not as national or regional representatives. The RRB approves the Procedural Rules, including the technical criteria used by the RRB in its application of the Radio Regulations. The RRB investigates instances of harmful interference and recommends solutions to such issues. The RRB advises Conferences on radio issues and appeals unresolved issues involving administrations to the WRC.

The Radio Regulations are normally adopted as Regulations by nations, even those not required by treaty to do so. In the United States, for example, the regulations are incorporated in 47 CFR Part 2-Frequency Allocations and Radio Treaty Matters; General Rules and Regulations.

The ITU-R Study Groups include the following:

Study Group 1:	Spectrum management
Study Group 3:	Radio wave propagation
Study Group 4:	Fixed-satellite service
Study Group 6:	Broadcasting service–sound (including vision and multimedia)
Study Group 7:	Science services (includes space operation, radio astronomy)
Study Group 8:	Mobile, radio determination, amateur and related satellite services
Study Group 9:	Fixed service (operations via terrestrial stations)
CCV:	Coordination Committee for Vocabulary (mainly with the IEC)
CPM:	Conference Preparatory Meeting (for the World Radiocommunication Conference (WRC)
SC:	Special Committee on Regulatory/Procedural Matters

Study Group 1 has several Working Parties that address specialized issues. The IEEE and the CISPR are also obviously involved in this area.

References

Breitenberg, Maureen, ed., 1989, *Directory of International and Regional Organizations Conducting Standards-Related Activities,* NIST Special Publication 767.

Cornish, Steven P., 2005, New ISO Policy Provides International Solutions to Market Needs, *ASTM Standardization News* 33(1): 38–41.

Cuschieri, Henry, 1998, Capital Hill Hearings Focus on International Standards as Barriers to Trade, *SES Standards Engineering* 50(3): 1–4.

Falk, Bernard H., Liew Mun Leong, and Pekka Tarjanne, 1997, World Trade Needs Worldwide Standards, *IEEE Power Engineering Review* 17(12): 38.

Grove, Jeff, 2005, U.S. Policy Makers Review Standards and Trade Issues, *ASTM Standardization News* 33(7): 24–26.

ISO/IEC Guide 59, 1994, *Code of good practice for standardization.*

Kirby, R. C., 1980, WARC plots spectrum use, *IEEE Spectrum* 17(2): 61–63.

Kitzantides, Frank K., 1992, Global Effects on U.S. Standardization Activities, *IEEE Power Engineering Review* 12(1): 13 –16.

Krechmer, Ken, 1996, Recommendations for the Global Information Highway: A Matter of Standards, *ACM StandardView* 4(1): 24–28.

Kushnier, Gary W., 2007, The Pacific Area Standards Congress (PASC), *ASTM Standardization News* 25(10): 19–21.

Lowell, Stephen, 1997, The Modern Day Archimedes: Using International Standards to Leverage World Markets, *SES Standards Engineering* 45(6): 1–6.

Mackay, Donald R., 2000, A Tripartite Program for the Expansion of World Trade, *SES Standards Engineering* 52(4): 7.

Mader, Donald A., and David C. Haataja, 1999, UL's Free Trade Policy: Transportable Conformity Assessment, *SES Standards Engineering* 51(2): 1–16.

Mason, John, 1980, Why Parts "Fit": The Role of the IEC, *IEEE Spectrum* 17(6): 42–44.

Matos, Fredrick, ed., 1985, *Spectrum Management and Engineering*, New York, IEEE Press.

Moore, James W., 1998, *Software Engineering Standards: A User's Roadmap*, Los Alamitos, CA, IEEE Press.

Nicolas, Florence, 1988, *Common Standards for Enterprises*, Commission of the European Communities, ECSC-EEC-EAEC, Brussels-Luxembourg, p. 44.

Office of Technology Assessment (OTA), 1992, *Global Standards: Building Blocks for the Future*, Washington DC, U.S. Congress, Office of Technology Assessment, p. 80.

Schonfeld, Raymond, and John Dobinson, 2004, Using International Standards: A Wake-Up Call to Regulators? *ASTM Standardization News* 32(1): 32–35.

Smoot, Oliver R., 2002, Breaking the Mold: A 21st Century Approach to Collaboration in the Standards Community, *SES Standards Engineering* 54(5): 1–7.

USTR, 2006, USTR Rejects Assertions that WTO TBT Agreement Designates Specific International SDOs, *ASTM Standardization News* 34(8): 6.

Voelcker, John, 1986, Helping Computers Communicate, *IEEE Spectrum* 23(3): 61–70.

Willingmyre, George T., 1997, Section 11, International Standards at the Crossroads, *ACM Standard View* 5(4): 190.

Zuckerman, Amy, 1997, *International Standards Desk Reference: Your Passport to World Markets, ISO 9000, CE Mark, QS-9000, SSM, ISO 14000, Q9000, American, European, and Global Standards Systems*, New York, American Management Association.

7

Regional Standards and Conformity Assessment

Introduction

The regional standards and conformity assessment organizations (bodies) are shown in the middle layer of Figure 6.1. It is readily seen that the emphasis here is on the European standards and conformity assessment bodies: the European Committee for Standardization (CEN), the European Committee for Electrotechnical Standardization (CENELEC), and the European Telecommunications Standards Institute (ETSI). The rationale for this emphasis on the European bodies is that they are the most developed and have assumed a large role in regional standardization activities. The European bodies have prompted the development of regional organizations around the world and served a model, at least in part, for other regional standards and conformity assessment bodies. CEN has working arrangements with the International Organization for Standardization (ISO), while CENELEC has close working arrangements with the International Electrotechnical Committee (IEC). ETSI has close working relationships with the International Telecommunications Union (ITU). The role of the European Union (EU) in these three bodies requires some understanding of the EU, referred to earlier as the European Economic Community (EEC), with the European Community (EC) as a major component.

European Union (EU)

The concept of a "United States of Europe" is attributed to Winston Churchill, the British wartime prime minister in 1946, just after the end of World War II (Pinder 2001). The goals of the formation were political (avoiding future wars) and economic (benefits of free trade). The first concrete step in the process was the European Coal and Steel Community (ECSC), a coalition of

France, West Germany, Belgium, Luxembourg, Italy, and the Netherlands in 1951. The ECSC removed trade barriers between its member nations for the commodities of coal, steel, iron, and scrap metal and was a successful model for the formation of the European Community (EC) in 1957. The member nations of the EC adopted the Single European Act in 1987. The goal of the act was the creation of a single market by the end of 1992 (frequently called Europe 1992 or EC '92). Some European nations formed the European Free Trade Agreement (EFTA), which participated only partially in the EC; some of those nations later joined the EC and most of the remainder joined with the EC to form the European Economic Area (EEA).

To form the single market in 1992, it was necessary for the EC to replace the national standards and conformity assessments of the member nations with regional (European) standards and conformity assessment schemes. The intense standards activity involved in Europe 1992 alarmed some in the United States who feared a "fortress Europe" that would hinder trade with the European Nations (Fitzgerald 1990). Some small U.S. exporters were hurt by European standards. One of several well-documented cases of EU harm to a small manufacturer is outlined by Helen Delaney (1996). The replacement of many national standards with single EC standards provided economies of scale for many large manufacturers. Thus, the activity in the EC on standards and conformity assessment woke U.S. managers to the importance of International Standards, which were the source of most (80–90 percent) of the European Norms (ENs). The "old approach" of harmonizing old standards and developing new standards was far too slow for the EC so a "New Approach" to standardization was developed.

The New Approach

In 1990, the Commission of the European Communities published the "Commission Green Paper on the Development of European Standardization Action for Faster Technological Integration in Europe." The Green Paper established the policy of handing off the development of EC standards to CEN, CENELEC, and ETSI. These bodies were staffed with full-time standards people as well as industry volunteers so that the work could proceed more quickly than in the usual Technical Committee environment. Annex 1 of the Green Paper listed Standardization Order Vouchers given to CEN and CENELEC since 1986 to help finance the work.

The principle of the New Approach is to include broad *essential requirements* in Directives that substitute for "Harmonized Standards" that are usually developed later to support the Directives. The Directives thus specify goals but do not prescribe the detailed methods of achieving those goals. Directives may be directly used for conformity assessment. It should be mentioned here that the European definition of the term "Harmonized Standard" is more restrictive than the ISO/IEC Guide 2 definition in Chapter 1 (Delaney and van de Zande 2000). In Euro parlance, "Harmonized Standards" are:

- Mandated by the European Commission
- Developed by European Bodies (CEN, CENELEC, ETSI)
- Address essential requirements of New Approach Directives
- Most important, published in the *Official Journal of the European Communities*, which notifies as their development

There are, of course, product standards that are not covered by the New Approach Directives.

Conformity Assessment: The Global Approach

Variations in conformity assessment from one Directive to another prompted the EC to issue a "Module Decision" in 1990. A variety-reduced set of conformity assessment procedures was adopted to provide economy and flexibility instead of a single one-size-fits-all approach. The modules represent a menu of the conformity assessment procedures that may be called out in directives. A regulation on CE marking was also promulgated. For many low-risk products, a manufacturer can (first party) attest (frequently called a Declaration of Conformity or DoC) that it meets all of the essential requirements of the applicable directive(s) by preparing a Technical File that shows how compliance is achieved and affixing the CE Mark to the product. For products where the risk is higher, the applicable directive may require third-party attestation (Certification) by a Notified Body (NB) or the use of NB oversight of the manufacturer's quality program (per EN 29000 standards that implement ISO 9000 standards).

For some regulated products, the Directive includes the conformity assessment criteria in the Directive itself, instead of by reference to the modules. Another aspect of the Global Approach is to promote the use of Government-to-Government Mutual Recognition Agreements (MRAs).

Table 7.1 shows the structure of the modules. Although eight possibilities are indicated, as many as twenty-four variations may be achievable according to Hanson (2005). An inspection of the modules in Table 7.1 shows the importance of the Notified Bodies and the European Quality System (QS) Standards.

Module A may be used alone to apply the CE Mark.

Module Aa is often used to obtain a voluntary Certification Mark to aid in marketing.

Module B is used with one of the other modules—C, D, E, and F—if required by Directive.

Modules G and H involve Notified Body participation, the latter for high-risk products that require "full quality assurance."

Notified Bodies (NBs) are Conformity Assessment Bodies that have been identified by the administrations of the Member states in the EU and

TABLE 7.1

Conformity Assessment Procedures in Community Legislation

A. (Internal control of production)	Conformity Assessment Procedures in Community Legislation				G. (unit verification)	H. (full quality assurance)
	B. (type examination)					
Manufacturer Keeps technical documentation at the disposal of national authorities Aa Intervention of notified body	Manufacturer submits to notified body — Technical documentation — Type Notified body — Ascertains conformity with essential requirements — Carries out tests, if necessary — Issues EC type-examination certificate				Manufacturer — Submits technical documentation	EN 29001 Manufacturer — Operates an approved quality system (QS) for design Notified body — Carries out surveillance of the QS — Verifies conformity of the design[1] — Issues EC design examination certificate[1]
	C. (conformity to type)	D. (production quality assurance)	E. (product quality assurance)	F. (product verification)		
		EN 29002	EN 29003			
A. Manufacturer — Declares conformity with essential requirements — Affixes the CE marking	Manufacturer — Declares conformity with approved type — Affixes the CE marking	Manufacturer — Operates an approved quality system (QS) for production and testing — Declares conformity with approved type — Affixes the CE marking	Manufacturer — Operates an approved quality system (QS) for inspection and testing — Declares conformity with approved type, or to essential requirements — Affixes the CE marking	Manufacturer — Declares conformity with approved type, or with essential requirements — Affixes the CE marking	Manufacturer — Submits product — Declares conformity — Affixes the CE marking	Manufacturer — Operates an approved QS for production and testing — Declares conformity — Affixes the CE marking
Aa Notified body — Tests on specific aspects of the product[1] — Product checks at random intervals[1]	Notified body — Tests on specific aspects of the product[1] — Product checks at random intervals[1]	Notified body — Approves the QS — Carries out surveillance of the QS	Notified body — Approves the QS — Carries out surveillance of the QS	Notified body — Verifies conformity — Issues certificate at conformity	Notified body — Verifies conformity with essential requirements — Issues certificate of conformity	Notified body — Carries out surveillance of the QS

[1] Supplementary requirements which may be used in specific Directives

notified to the EU and the other Member States. NBs must meet accreditation standards and are roughly analogous to the Nationally Recognized Testing Laboratories (NRTLs) in the United States (Chapter 2). The identities of Notified Bodies are published in the *Official Journal of the European Communities.* "Competent Bodies" supplement Notified Bodies for conformity assessment in the EMC Directive.

Surveillance Authorities are Member State agencies (national authorities) that monitor the use of the CE Mark for products that have been put into use. If a surveillance authority discovers a CE Marked product on the market that does *not* meet the essential requirements of the appropriate Directive(s) and the Harmonized Standards, if any, it takes corrective action within the EU to mitigate any resulting hazard.

Anatomy of a Directive: The Low-Voltage Directive

One of the earliest and most important EC Directives was the Low-Voltage Directive:

> COUNCIL DIRECTIVE of 19 February 1973 "**on the harmonization of the laws of member states relating to electrical equipment designed for use within certain voltage limits (73/23/EEC)**."

The Low-Voltage Directive (LVD for brevity) had some issues that were documented in a commentary (Winkler et al. 1975). The LVD did not start life as a New Approach Directive but was converted to one by Council Directive 93/68/EEC (30 August 1993) that prescribed the use of the modules and CE marking for the LVD and certain other Directives.

Since the LVD had been modified several times, it was "reissued" in a "codified version" as Directive 2006/95/EC, which is used here to illustrate the anatomy of a directive.

Preamble: Justifications

Article 1	(Scope) "… equipment rated from 50 to 1000 V for ac and 75-1500 V dc."
Article 2	Member states to address electrical safety (references safety objective in Annex I).
Article 3	Member states to assure free movement of compliant electrical products.
Article 4	Member states not to lay down stricter requirements.
Article 5	Member states to assure that electrical equipment complies with harmonized standards.
Article 6	Use of other standards if harmonized standards are not available.

Article 7	Use of standards in force in Member states if others (5) & (6) are not available.
Article 8	CE Marking is mandated and reference to conformity assessment in Annex IV.
Article 9	Member state to notify other Member states if use of equipment is prohibited (an example of the so-called Safeguard clause in Legislation).
Article 10	CE conformity marking in Annex III is mandated.
Article 11	Member states to communicate Notified Bodies to Other Member states and The Commission.
Article 12	Directive not applicable for electrical equipment for export to third countries.
Article 13	Member states to communicate to Commission texts of adopted national laws.
Article 14	Directive 73/23/EEC is repealed and references to it deemed to be made to this Directive.
Article 15	Directive to come into force on 20th day following publication in the *Official Journal of the European Union*.
Article 16	This directive is addressed to the Member States.

Signed (for the European Parliament) and (for the Council).

Annex I	Principal Elements of the Safety Objectives for Electrical Equipment Designed For Use within Certain Voltage Limits (now called "essential requirements").
Annex II	Equipment and Phenomena outside the Scope of this Directive.
Annex III	CE Conformity Marking and EC Declaration of Conformity.
Annex IV	Internal Production Control (*Selection of the "A" module for conformity Assessment in this Directive, emphasis added*).

The scope and importance of the Low Voltage Directive is substantiated by the fact that more than 600 "harmonized standards" have been developed to support it. The Machinery Directive is another broad document. Some Directives are much more specialized, for example, the EU Active Implantable Medical Devices Directive. The National Institute of Standards and Technology has published a series of guides for the major Directives that are listed in Appendix 5.

CEN/CENELEC: The Joint European Standards Institutions

CEN and CENELEC, like the ISO and the IEC, share many basic principles. Both are situated in Brussels, Belgium. The main goals of CEN and CENELEC, established in 1976 and 1973, respectively, are to provide European standards for the internal market and to promote European competitiveness in the rest of the world. As with the International Standards bodies, the main technical work is done in Technical Committees and Working Groups. The Secretariats in Europe are usually allocated to the same Member States that hold corresponding Secretariats in ISO and IEC (Nicolas 1988). Most CEN and CENELEC standards are transposed ISO and IEC standards, respectively. Both organizations have programming committees that assure that the standards developed reflect the priorities of the EU. Except in exceptional circumstances, CEN and CENELEC members have agreed not to publish national standards when a corresponding European standard is being prepared (the so-called standstill agreement). European prestandards (ENVs) are exempt from the standstill agreement.

"Harmonized standards" are approved on the basis of a "qualified majority" rather than the classical consensus methods in ISO and IEC (CEN/CENELEC 2002). The criterion for approval takes into account the number of countries voting against a proposal, the number of abstaining countries, and the number of negative votes. There is an appeal procedure that seeks fairness in the decision. The CEN and CENELEC deliverables include the following:

EN—European Norms

HD—Harmonization documents

ENV—European prestandards

CEN and CENELEC Workshop Agreements (CWAs)

European Norms must be taken over (adopted) by Member States according to a schedule for each document. For example, Amendment 3 of EN 60950, a CENELEC document issued in 1995, required Member States to follow the schedule:

dor 1994-12-06

doa 1995-04-01

dop 1996-01-01

dow 1997-01-01

where

dor is the date of *ratification* by the CENELEC Technical Board

doa is the latest date for *announcement* of the EN/HD at the national level

dop is the latest date by which an EN must be implemented at the national level by *publication* (or endorsement) of an identical national standard

dow is the date by which conflicting national standards must be *withdrawn*

The National Member adoption of HDs is somewhat relaxed as long as the corresponding national standards are "equivalent" to HDs. The schedule and rules apply only for the EFTA countries that voted in favor of the HD. The transposition of an EN to a national standard is indicated only by the designation; EN 60950, for example, becomes BS EN 60950 in Britain.

The ENVs and CWAs are temporary documents for quick reaction to new technologies. They must be converted to ENs or HDs within five years or retired. Additional special documents are discussed in the descriptions of CEN and CENELEC.

The three official languages of CEN/CENELEC are English, French, and German. The original text of an ENV or SWA may initially be published in only one of the official languages. Like ISO and IEC, CEN and CENELEC are governed by a common set of internal regulations.

European Committee For Standardization (CEN)

A simplified organization chart for CEN is shown in Figure 7.1. The "Counsellors" are the EU and EFTA bodies such as the Director General, Enterprise, and Industry. The EU Director General provides standardization mandates and some of the funding for the CEN. The parts of the organization in the center column generally report to the parts of the organization immediately above them. The General Assembly (AG) is

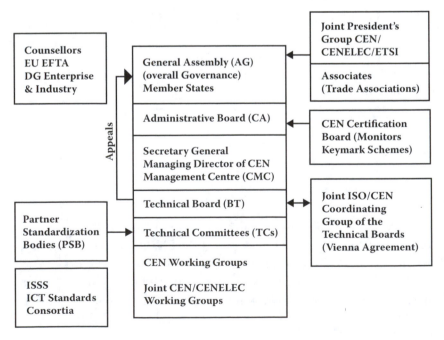

FIGURE 7.1
Simplified organization of CEN.

composed of representatives from the National Standards Bodies (NSBs) of EU Member States.

The AG sets the overall policies of CEN, including internal and external matters, and handles appeals lodged with the Technical Board. The General Assembly creates and maintains the statutes that govern CEN and delegates certain responsibilities to the Administrative Board. The CEN Certification (Conformity Assessment) Board reports to the Administrative Board as does the Secretary General, which manages the CEN Management Center (CMC). The CMC provides most "Secretariat" functions in CEN. The CEN Technical Board is the analog of the Technical Management Board (TMB) in ISO and is part of the Joint ISO/CEN Coordinating Group of the Technical Boards pursuant to the Vienna Agreement. The Technical Board decides which proposals to pursue and assigns the work to an appropriate Technical Committee or creates a new Technical Committee (TC) to handle it. The various Technical Committees report to the Technical Board and set the standardization goals for the CEN Working Groups, which is where the actual work of making standards is done. Some of the Working Groups are Joint CEN/CENELEC Working Groups. The Partner Standardization Bodies (PSBs) provide outside input to the Technical Committees. A PSB is a National Standards Body, an ISO member, and unlikely to become a CEN National Member. The host countries of the PSBs commit to using CEN standards.

Historically, those outside the EU could not readily see how the CEN organization operates or have substantial input to its standardization work except by working through the ISO. CEN has opened up somewhat in recent years, especially in the fast moving area of information and communication technology (ICT). CEN recently created an Information Society Standardization System (ISSS) that issues a special form of a prestandard document called a CEN Workshop Agreement (CWA). This arrangement was prompted by the fact that many ICT standards are now being produced by Consortia rather than the usual Technical Committee process (Rada and Ketchell 2000). Unlike the usual Technical Committees, outsiders can participate in developing CWAs. The ISSS acts as a kind of specialized secretariat for CWA support. CEN has published a list of "accepted consortia" for the purpose of developing CWAs. Accepted consortia must be legitimate active consortia that have *international* members that are not pushing single national standards. CEN's survey list of Fora and Consortia include the 1394TA, the ACM, the Bluetooth Consortium, the DSL Forum, the DVD Forum, and W3C.

Many strong connections unify the operations of CEN, CENELEC, and ETSI, including the Joint Presidents Group, which works with the General Assemblies. The Associate Members are Trade Associations and others whose views are inputs to the General Assemblies. The CEN Certification Board monitors the CEN and "Keymark" certification schemes, which are operated by many Certifiers in the Member Nations.

The Vienna Agreement

The CEN Administrative Board and the ISO Executive Board endorsed the *Vienna Agreement* (VA) in 1991. The VA has been modified several times in view of operating experience with it. Objectives of the VA include reduced duplication of effort, openness, recognition of regulatory needs (on CEN's part), and global relevance in accordance with ISO principles. Either body can take over standards of the other body and participate in the work of the other body. Common meetings are to be avoided, but meetings may take place in tandem at the same location. Parallel voting is a major issue involved in the VA to speed up standard setting. The VA includes rather detailed procedures for handling votes in the two bodies that do not agree and for the resynchronization of development and maintenance of standards. The lead body is usually CEN for standards efforts in support of EU Directives. The CEN Technical Board (BT) along with the ISO Technical Management Board (TMB) monitors the application of the VA. In the event that the EU mandates a standard, the published standards are identical except that an Annex Z is included in the CEN publication but not in the ISO document.

European Committee for Electrotechnical Standardization (CENELEC)

CENELEC was born to parent organizations that had been active in standardization in Europe since the early 1950s. Like ISO, it is a nonprofit technical organization made up of the Electrotechnical Committees of the EU Member States, which are the same as the National Committees of the IEC. CENELEC members represent the interests of manufacturers, consumers, users trade associations, and utilities and network operators, among others. Like ISO, CENELEC aims to promote the use of IEC/CENELEC standards throughout Europe and abroad.

The sources of CENELEC standards are varied. They include IEC standards (accounting for nearly 90 percent of CENELEC standards), harmonized standards of Member States, new publications of Technical Societies, the outputs of CENELEC Workshop Agreements, and any other suitable document that the EU and Industry need. CENELEC has developed more stringent standards in some cases where the content of a corresponding IEC standard was considered to be inadequate for Europe (as in some safety standards, for example).

The organization of CENELEC is shown in the simplified organization chart in Figure 7.2. The General Assembly (AG) is the governing body. The General Assembly receives input from the nonvoting Associates and also from the NAFTA Secretariats. The President of CENELEC oversees the Administrative Board (CA), which develops policies and monitors the operation of the General Secretariat. The General Secretariat is headed by the Secretary General (GS). The Secretary General supports the General

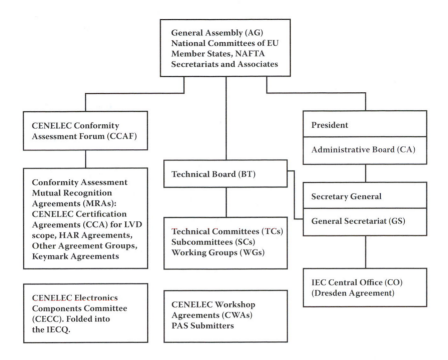

FIGURE 7.2
Simplified organization of CENELEC.

Assembly and the Administrative Board by supplying secretariat services to the Technical Board (BT) and the Technical Committees (TCs). The TCs include the Subcommittees (SCs) and Working Groups (WGs). The TCs, SCs, and WGs include representatives of the Member State electrotechnical committees. The General Secretariat also interfaces with the IEC Central Office (CO) pursuant to the Dresden Agreement.

The CENELEC Conformity Assessment Forum (CCAF) organizes the area of conformity assessment for electrotechnical subjects in Europe. The CCAF replaced the former ELSECOM Committee in conformity assessment matters. The CCAF also serves as a coordinator of conformity assessment issues between the IEC, the EU Commission, CEN, ETSI, and the EFTA Secretariat. Conformity assessment is diversified, as elsewhere, by the use of Mutual Recognition Agreements (MRAs) between Member State agencies and also by a number of CENELEC Certification Agreements (CCAs) for issues falling under the Low-Voltage Directive. Other agreement groups are devoted to HAR issues on cables and cords, the ENEC Agreement for the safety of IT equipment pursuant to EN 60950, and EMCRAFT for EMC testing in broad fields of products. The CCAF also monitors the Keymark agreement groups, a widely used "quality" mark for consumer products in Europe. A separate CENELEC Electronics Component Committee (CECC), which duplicated the function of the IEC Quality Assessment System for Electronic Components (IECQ), has been merged into the IECQ.

The Dresden Agreement

The *Dresden Agreement* between the IEC and CENELEC grew out of an earlier cooperative agreement called the Lugano Agreement that was made in 1991. The Dresden Agreement, concluded in 1996, is analogous to the ISO/CEN Vienna Agreement. The Dresden Agreement calls for the common planning of new work and the parallel voting on Committee Drafts for Vote (CDVs) and the Final Drafts of International Standards (FDISs), with some exceptions. The procedures are regulated by the ISO/IEC Directives as well as the CEN/CENELEC Regulations. The CENELEC General Secretariat provides the German text of FDIS documents prior to the vote on them. A rather complex procedure is involved for handling votes at the Ratification stage, that is, one body votes positive while the other votes negative, etc. A provision exists for the submission by the Technical Board of European EN and HDs to the IEC as FDIS documents!

European Telecommunications Standards Institute (ETSI): The Convergence of Information and Communication Technologies

The 1980s saw the deregulation of telecommunications with the breakup of the Bell System. That wave of deregulation swept over Europe and the rest of the world. Later in that decade it became apparent that telecommunication was being computerized and that new standards and regulatory regimes would be needed to cope with these developments. Specifically, the effects of Europe 1992 on these issues came into focus (Rankine 1990). The U.S. concern was that it was being left out of the European standardization developments. As of 1990, some parts of Europe did not enjoy touch-tone service (Stix 1990). Europe was concerned that it might fall behind the United States and Japan in ICT standardization if it did not act aggressively. As a result, the European Community developed far-reaching policies on the development of telecommunications in Europe (Ungerer and Costello 1988). It was recognized that a "Single Europe" depended on the creation of a new standards and conformity infrastructure for ICT. M. H. Sherif (1998) has provided a more recent paper on convergence surveying ETSI's role and the division of labor in standard setting among the ITU, IETF, and other organizations.

Because the European development of telecom standards was too slow, the EC created ETSI in 1988 to speed up the preparation of the badly needed standards. ETSI is an independent non-profit standards body. Like its sister organizations, ETSI has the European Commission and the EFTA Secretariat as "Counselors." Some full-time paid staff (including Project Teams) in addition to the usual Technical Committees helped accelerate the standards setting process. ETSI is located in Sophia Antipolis in France, unlike its sister organizations in Brussels. ETSI is relatively open compared to CEN and CENELEC. Like the ITU, is has a wide membership that participates in developing ETS standards, some of which are "adoptions" of ITU standards. At last report, ETSI had approximately 700 members from some 55 countries. ETSI may thus be considered a de facto International Standards body, like a mini-ITU. Unlike the ITU, participants do not have to go through a "state department" as shown in Figure 6.1. Like its sister organizations in Brussels— CEN and CENELEC—the National Standards Bodies in ETSI operate on a one-vote-per-country basis to produce draft European Telecommunications Standards (ETSs). Before ETSI issues a standard, it is subject to weighted voting. ETSs are voluntary standards, like the ITU "Recommendations," some of which are transposed to ETSI standards. The process follows the New Approach system (Ask 1995). ETSI also produces Europe's Technical Basis for Regulations (TBRs) where it is known in advance that the TBR will be converted by the EU Technical Recommendations Committee (TRAC) to a Common Technical Regulation (CTR). The process is roughly analogous to the way U.S. government agencies include voluntary standards into regulations (Incorporation by Reference). There are far too many interconnections between ETSI and other standards organizations to even hint at in Figure 6.1. ETSI is a major player in the global collaboration on ICT standards (Habara 1994). Habara's paper shows how a truly remarkable collaboration of regional and national telecom standards bodies have worked to advance ICT standards within the context of the ITU. Habara also points out that the term "regional standards organization" in the ICT context is "somewhat imprecise," so a new term, the "participating standards organization" (PSO), was coined to the replace it. A recent paper by Imre Varga et al. (2006) shows how ETSI, the ITU, and the IETF (and industry) cooperate to develop modern ICT standards. The development and use of the "GSM" ETSI standards for Europe proved to be far superior to the variety of cellular standards developed in the United States (Brull 1998).

Conformity Assessment and Interoperability

It has sometimes been mistakenly asserted that equipment that conforms to a standard will interoperate with other equipment that conforms to that standard. Although conformance to a standard might be considered a *necessary* condition for interoperability, it is not a *sufficient* one except in the simplest of cases. Protocol standards are designed to realize interoperability, like the IEEE 802 standards. ETSI has a Protocol Testing and Competence Center (PTCC) that supports the development by clients of testing standards. ETSI conformance testing is based on ISO/IEC 9646 (a JTC 1 standard).

Conformance testing involves the use of "host" equipment. The host equipment implements a standard, uses standard interfaces to supply protocol signals to the equipment under test, and then measures the EUT's responses. In addition, ETSI has a Plugtest™ Service that empirically tests equipment for interoperability (end-to-end functionality with two or more communicating systems).

The relative merits of conformity testing and interoperability testing have been widely debated. ETSI has adopted the position that the coordinated use of both is the most reliable. Certification of conformity and interoperability testing is a useful tool that ETSI delegates to independent organizations. ETSI has a guide for Specification Languages that are used in some protocols. The relative openness of ETSI as compared with the CEN/CENELEC system is a win-win situation for the EU, which gains ICT input, and the members that gain access to standards setting in the EU and ITU. More details can be found at www.etsi.org.

ETSI and Intellectual Property Rights (IPR)

In its earlier days, ETSI was involved in a "celebrated" dispute over the inclusion of IPR (mostly patent rights) in its standards. This topic will be addressed in Chapter 10.

Regional Standards and Conformity Assessment in Other Areas

As shown in Appendix 7, there are more than fifty Regional standards organizations. Space limitations permit the treatment of only a few of the most important ones here.

Asia Pacific Telecommunity (APT)

The Asia Pacific Telecommunity, established in 1979, is shown on the right-hand side of Figure 6.1. The APT ICT Task Force was established by the United Nations Economic and Social Commission for Asia and the Pacific (UN-ESCAP), using the offices of the ITU. The ICT Task Force Regional Network in Asia resulted. The rationale for this activity is to promote trade and development since ICT facilities are necessary for developing nations. A main thrust of this activity is to mitigate the so-called digital divide, the lack of developing nations' access to ICT. The first preparatory meeting for setting up the ICT Task Force Regional Network in Asia was held in February 2002 in Bangkok, where the Secretariat is located. Several workshops on Next Generation Networks and preparation activities for participation in the World Radio Conference (WRC) have also been held over the years. An APT Regional Workshop on Standardization was held in July

2007. More information on the APT is available at the Secretariat Web page: www.aptsec.org.

Inter-American Telecommunication Commission (CITEL)

As shown in the right-hand side of Figure 6.1, CITEL is a creature of the Organization of American States (OAS) treaty, hence the U.S. State Department is involved for U.S. participants. CITEL is based in Washington DC and covers the Americas. CITEL includes some 30 Member States and 200 Associate Members. CITEL is committed to work within the frameworks of International Bodies including the World Trade Organization (WTO/TBT), ITU, ISO/IEC, and appropriate regional bodies. CITEL is roughly modeled after ETSI except that the emphasis is on Conformity Assessment rather than the setting of standards.

The March 1988 Declaration of Quito spelled out the goals of CITEL, including the achievement of social and economic goals. The declaration confirms the primacy of CITEL in the development of the region's telecommunications sector and emphasizes meeting the needs of less developed areas, especially in the areas of telemedicine, distance learning, and access to the Internet. The centerpiece of CITEL in this work is the Inter-American Mutual Agreement for Conformity Assessment of Telecommunications Equipment (October 1999). The MRA is based on confidence building in the use of Conformity Assessment Bodies using ISO/IEC Guides (see Chapter 2) and the use of Accreditation standards (Chapter 3). The MRA calls for the Designation of CABs and the "verification" of Conformity Assessment Bodies pursuant to appropriate International Standards. Unlike the situation in the EU, the MRA preserves the regulatory authority of its Members. One can see the influence of the EU's Notified Bodies concept in the MRA.

Since the Members of the OAS are contiguous "states," it should be easier to work through some of the regional issues involved than in some other regional bodies. The Alliance of Telecommunications Industry Solution (ATIS) and the Telecommunications Industry Alliance (TIA) are major U.S. organizations active in CITEL (more on them in Chapter 8). More information on CITEL appears on the Web site www.citel.oas.org.

Asia-Pacific Economic Cooperation (APEC) and Pacific Area Standards Congress (PASC)

The APEC includes member nations along the Pacific Rim, a large but rather disparate economic region. APEC headquarters is in Singapore and includes twenty-one member economies, including the United States, China, Australia and New Zealand. Established in 1989, APEC's goals are free and open trade for developed countries by 2010 and by 2020 for developing countries. APEC's role in standards and conformity assessment was highlighted in the National Research Council's report *Standards, Conformity*

Assessment, and Trade (NRC 1995). APEC has developed a "Telecom MRA," an Electrical Equipment MRA, and other arrangements in the region including a Subcommittee on Standards and Conformance (APEC SCSC).

The APEC SCSC recognizes a number of Conformity Assessment organizations as follows:

Asia-Pacific Laboratory Accreditation Cooperation (APLAC)

Asia-Pacific Legal Metrology Forum (APLMF)

Asia-Pacific Metrology Program (APMP)

Pacific Accreditation Cooperation (PAC)

Pacific Area Standards Congress (PASC)

The Pacific Area Standards Congress, organized in 1973, includes many of the APEC "members" National Standards Bodies. PASC has developed programs for the use of the International Standards of the ISO and IEC in Conformity Assessment in that region. PASC is one of five Regional Bodies recognized by the APEC SCSC. ANSI is the U.S. representative to PASC. Gary Kushnier (1997) of ANSI has provided some insight into the structure and operation of PASC. Kushnier points out that PASC can develop a unified voice at the ISO and IEC levels. PASC has also undertaken helping Asia-Pacific governments meet standardization needs and has asserted that the Asia-Pacific area does not have to set up "regional standards bodies" like the CEN/CENELEC/ETSI structure in the EU. NIST, along with some U.S. safety and code agencies, sponsored a weeklong workshop "Electrical Safety Systems in the Asia-Pacific" during March 2002. More information is available on the PASC Web site at www.pascnet.org/.

North America Free Trade Agreement (NAFTA) and Council for Harmonization of Electrotechnical Standards of the Nations of the Americas (CANENA)

The North America Free Trade Agreement between Mexico, Canada, and the United States was preceded by the U.S.–Canada Free Trade Agreement (CFTA). Both agreements recognized the role that standards and conformity assessment played in trade between the nations. The Canadian Standards Association (CSA) and Underwriters Laboratories (UL) had already commenced the process of harmonizing U.S. and Canadian standards in broad electrical areas. The same concepts were extended to include Mexico in the NAFTA. In most cases, IEC standards were the basis for the harmonization of electrical standards. Several binational and even some trinational standards were processed. The private sector work was sponsored by the Standards Council of Canada (SCC), the Mexican national chamber for the manufacturing industries (CANACINTRA), and ANSI (IEEE 1992). More details appear in the Office of NAFTA and Inter-American Affairs documents of the International Trade Administration, Market Access and Compliance.

CANENA was born in 1992 following the North American Free Trade Agreement. As the name implies, CANENA is not a Standards Developer but promotes the *harmonization* of national electrical codes and standards in North America. Mexico had adopted a national code similar to the National Electrical Code of the United States. CANENA provided the organizational framework for the development of some of the binational and trinational standards mentioned earlier. Conformity Assessment test methods are also included in the purview of CANENA. Technical Harmonization Committees (THCs) correspond to Technical Committees (TCs) of the IEC. A THC cannot be established unless at least two of the countries request it. Consensus in a THC follows the definition of ISO/IEC Guide 2. CANENA has signed cooperation agreements with both IEC and CENELEC.

Free Trade Area of the Americas (FTAA) and Pan American Standards Commission (COPANT)

The FTAA was born at the Summit of the Americas in 1994. The objective of the FTAA is to, among other things, eliminate trade barriers in the region. NIST supports the U.S. Trade Representative in its negotiations. One of the Working Groups involved addresses the issues of the TBT Agreement. Carmina Londono (1997) has presented a review of the FTAA as well as some other regional trade areas not covered here including the *Andean Pact, the Central America Common Market (CACM), the Caribbean Common Market (CARISOM), the G3 Group of 3*, and *Mercosur.*

COPANT was founded thirty-five years ago as an association of National Standards Bodies in the countries of Latin America to promote the development of technical standardization and related activities in member countries (Breitenberg 1989). The official language of COPANT is Spanish. The IEC has signed a Cooperation Agreement with COPANT and COPANT recently met with the PASC. NIST supports ANSI in COPANT activities (NIST GSIG Western Hemisphere Standards Related Activities).

Association of Southeast Asian Nations (ASEAN)

ASEAN was formed in Bangkok by five countries: Indonesia, Malaysia, Philippines, Singapore, and Thailand. The goals of the association are to speed economic growth as well as social progress and cultural development. The achievement of regional peace and stability through the rule of law and the United Nations principles were also goals of ASEAN. The ASEAN Economic Community, as its name implies, is to develop a single market, patterned in many respects after the single-market model of the European Union. Among the integration objectives was the development of an MRA-based infrastructure, the Telecommunications Regulators Council Sectoral

MTA (ATRC–MRA). The ASEAN developed further by the inclusion of China, Japan, and the Republic of Korea, called the ASEAN Plus Three program. Many members of ASEAN participate in the Asian Pacific Economic Cooperation (APEC) and other regional groups. A free trade agreement between ASEAN and Australia/New Zealand is also being developed.

A distinguishing feature of ASEAN is the robustness of its arrangements in standards and conformity assessment. ASEAN embraced the development of ICT to advance economic objectives in the region and to mitigate the digital divide. An ASEAN Memorandum of Understanding (MoU) with China aims at the development of ICT infrastructure in the region.

The ASEAN Cooperation on Standards and Conformance to Facilitate Trade in the Region recognizes that differing Conformity Assessment methods can impede trade as much as differing standards. ASEAN has implemented a Consultative Committee on Standards and Quality (ACCSQ) to guide the harmonization processes. In addition to several MRAs, the members have harmonized its standards with many International Standards for safety and electromagnetic compatibility (EMC). ASEAN also has analogs of the Notified Body arrangements in the EU. The ASEAN Policy Guideline on Standards and Conformance mandates the terminology in ISO/IEC Guide 2 (2004 edition) and ISO/IEC Standard 17000 (2004 edition) and prescribes the policies on regional MRAs. Moreover, the policy recommends the harmonization of existing national standards via the current edition of *ISO/IEC Guide 21: Adoption of International Standards as Regional or National Standards* (emphasis added). The policy further encourages observance of the WTO-TBT guides, including the discouragement of adopting "prescriptive standards." The policy even mentions the EU institution of "Post Market Surveillance." The correspondence of ASEAN standards and conformity assessment is so complete that ASEAN and EU are negotiating to form a *region-to-region* approach, the Trans-Regional EU-ASIAN Trade Initiative (TREAT). There is no indication that ASEAN will aim for the political unity of the EU. ASEAN and the PASC can probably provide the APEC with the infrastructure in standards and conformity assessment that it needs. More information on ASEAN is available at its homepage: www.aseanec.org.

Concluding Remarks

The importance and primacy of International Standards and Guides for use at the regional and national levels has been demonstrated. A developing region probably does better to start small and grow by absorption rather than casting too big a blanket to start.

In addition to the stimulus that it provided for the formation of other economic regions in the world, the EU served as an example, at least in part, for the organization of those regions. The IEC and CENELEC as well as Japan have substantial outreach programs to the regional standards bodies, a form of "industrial policy" that the United States would do well to emulate. Because of space limitations, there is no coverage here of other economic regions such as Mercosur in South America.

References

Ask, Frede, 1995, Interoperability and Intellectual Property, in *Standards Policy for Information Infrastructure*, edited by Brian Kahn and Janet Abbate, Cambridge, The MIT Press, pp. 405–409.

Breitenberg, Maureen, ed., 1989, *Directory of International and Regional Organizations Conducting Standards-Related Activities*, NIST Special Publication 767, p. 365.

Brull, Steven B., Neil Gross, and Catherine Young, 1998, Cell Phones: Europe Made the Right Call, *Business Week*, September 7, pp. 107, 110.

CEN/CENELEC 2002, *CEN/CENELEC Internal Regulations—Part 2: Common Rules for Standards Work* (para. 11.2.1.1).

Delaney, Helen, 1996, Stacking the Deck in Europe, *ASTM Standardization News* 24(8): 34–37.

Delaney, Helen, and Rene van de Zande, 2000, *A Guide to EU Standards and Conformity Assessment*, NIST Special Publication 951.

Fitzgerald, Karen, 1990, Global Standards: Facilitators or barriers? *IEEE Spectrum* 27(6): 44–46.

Habara, Kohei, 1994, Cooperation in Standardization, *IEEE Communications Magazine* 32(1): 78–84.

Hanson, David, 2005, *CE Marking, Product Standards and World Trade*, Cheltenham, UK, Edward Elgar, p. 54.

IEEE, 1992, Canada, Mexico, U.S. Promote Conformity, *IEEE Power Engineering Review* 12(7): 22.

Kushnier, Gary W., 1997, The Pacific Area Standards Congress (PASC), *ASTM Standardization News* 25(10): 19–21.

Londono, Carmina, 1997, A Free Trade Area of the Americas by the Year 2005? *ASTM Standardization News* 25(6): 28–34.

National Research Council (NRC), 1995, *Standards, Conformity Assessment, and Trade: Into the 21st Century*, Washington DC, National Academy Press, p. 154.

Nicolas, Florence (with Jacques Repussard), 1988, Common Standards for Enterprises, Brussels and Luxembourg, Commission of the European Communities (E), ECSC-EEC-EAEC.

Pinder, John, 2001, *The European Union*, Oxford, UK, Oxford University Press, p. 4.

Rada, Ray, and John Ketchell, 2000, Standardizing the European Information Society, *Communications of the ACM* 43(3): 21–25.

Rankine, L. John, 1990, The European Community: Will Standards Open or Close the Market in 1992? *Crossroads of Information Technology Standards*, Washington DC, National Academy Press, pp. 15–16.

Sherif, Mostafa Hashem, 1998, Convergence: A New Perspective for Standards, *IEEE Communications Magazine* 36(1): 110–111.

Stix, Gary, 1990, Telecommunication, *IEEE Spectrum* 27(6): 25–26.

Ungerer, Herbert, and Nicholas P. Costello, 1988, *Telecommunications in Europe*, Luxembourg, Commission of the European Communities.

Varga, Imre, Rosario Drogo De Iacova, and Paolo Usai, 2006, Standardization of the AMR Wideband Speech Codec in 3GPP and ITU-T, *IEEE Communications Magazine* 44(5): 66–73.

Winkler, Rudolf, J. Cassassolles, and D. Verdiani, 1975, *Commentary on the "Low Voltage Directive" of the Council of the European Communities of February 1973*, Berlin, VDE-Verlag GmbH. (English language edition edited by W. L. Leicester—Technical Help to Exporters.)

8

National Standards, Conformity Assessment, and Accreditation Bodies

Introduction: Do Unique National Standards Have a Future?

In light of the wholesale taking over (adoption) of International Standards to replace traditional national standards, is there a future for national standards? It seems clear that unique national standards will be of diminished importance in the future. The importance of International Standards has been repeatedly emphasized. Stephen Oksala (1996) points out that the connection between national standards and legislation seems to assure a continuing place for national standards. This point of view is reinforced by Shelley Waters Deppa (1997), who documents a possible exception to the "International only" rule. One rationale for unique national standards is that some of them may be those adopted by states, for example, where International Standards may not be a good fit. A firm that adheres to a unique national standard, where an International Standard is available, runs the risk of loosing out to competitors (both national and international). Where a nation (e.g., United States) leads in a given technology (e.g., computers), it is obviously important that national standards be created that can easily be transposed to International Standards. The National Standards Bodies (NSBs) still have important functions to perform, in addition to processing national standards and accrediting Conformity Assessment Bodies (CABs), as discussed in the following.

Standards, Conformity Assessment, and Accreditation in the United States

As described in Chapter 6 (Figure 6.1), the American National Standards Institute (ANSI) plays the key role of the National Standards Body (NSB) for approving American National Standards (ANSs) and interfacing with the ISO and IEC. Information and communication technology (ICT) bodies, while included in the ANSI umbrella in most cases, are conveniently covered later in this chapter. The federal standards and conformity assessment bodies (agencies) are discussed in Chapter 11. Toth (1991) provides, in NIST Special Publication 806, a broad summary of the many standards organizations (both government and private sector) in the United States. Hyer (2001) provides, in NIST Special Publication 903, a summary of the private sector U.S. Product Certification Programs, which does not exclude Registrars and Certification Bodies for Quality Management Systems and Environmental Management Systems. As indicated in the literature, the top twenty Standards Developers in the United States produce approximately 90 percent of the private sector standards. Among the largest (non-ICT) standards bodies in the United States are (abbreviations are in parentheses ( ), categories are in brackets [ ]):

Aerospace Industries Association (AIA) [Trade Association]

American Petroleum Institute (API) [Trade Association]

American Society of Mechanical Engineers-International (ASME) [Engineering/Professional]

American Society for Testing and Materials International (ASTMI) [Membership Organization]

Electronic Industries Alliance* (EIA) [Alliance of Trade Associations],

Institute of Electrical and Electronics Engineers* (IEEE) [Scientific–Professional Society]

National Electrical Manufacturer's Association (NEMA) [Trade Association]

National Fire Protection Association (NFPA) [Membership Organization]

Society of Automotive Engineers (SAE) [Professional Society]

Underwriters Laboratories (UL) [Independent SDO and CAB]

Coverage of even a significant fraction of these bodies is clearly impractical in the broad coverage of this book. Accordingly, a few significant U.S. bodies are discussed in the following and the U.S. system is then compared with those of several NSBs and subordinate SDOs (Standards Developing Organizations) and CABs outside the United States.

The American National Standards Institute (ANSI)

ANSI is frequently called a "federation." It is a private, nonprofit membership organization that includes most of the significant national standards

---

* Also appears in the "ICT bodies" listed later in this chapter. Most Consortia of interest here are involved in the ICT field.

and conformity assessment bodies in the United States. ANSI itself does not develop standards; it coordinates the development of Standards (including the gatekeeper function described in Chapter 4). ANSI's headquarters are in Washington DC with most of its infrastructure located in New York City. Although ANSI is focused on the private sector, it also plays an important role in cooperation with the federal sector, particularly with the National Institute of Standards and Technology (NIST). As mentioned in Chapter 4, ANSI accredits Standards Developing Organizations (SDOs) that operate under the ANSI Essential Requirements, stressing openness, balance, and consensus as part of due process. In addition, because of the wide public review that candidate ANSI standards go through, they frequently become candidates themselves for adoption by federal, state, and local authorities, thus becoming Regulations and Codes.

Referring to the ANSI organization chart in Figure 8.1, ANSI sponsors a series of *forums* for information sharing by interest groups including Company Member, Consumer Interest, Government Member, and Organizational Member Forums. In addition to networking on issues of common interest, the Company Member Forum (CMF) provides early warning of issues likely to affect company members. Emerging issues are also aired at the CMF. The Government Member Forum (GMF) works with the Interagency Committee on Standards Policy (ICSP) to improve cooperation between standards groups in the public and private sectors. The Consumer Interest Forum aims to improve the use of standards and conformity assessment to assist consumers. On the other side of the organization chart are the *Standards Panels*. The Standards Panels focus on emerging requirements for new standards such as the Homeland Security Standards Panel, the Nanotechnology Standards Panel, the Healthcare Information Technology Standards Panel, the Identity Theft Prevention and Identity Management Standards Panel, and the Biofuels Standards Panel. The Healthcare Information Technology Standards Panel is a ten-year program with funding from the U.S. Department of Health and Human Services. Old standards panels die and new ones are born as conditions change.

The next layer down in Figure 8.1 includes *policy committees*, such as the Conformity Assessment Policy Committee (CAPC). The CAPC provides policy input to the International Conformity Assessment Committee, the Accreditation Committee for Product Certifiers, and the Personnel Certification Accreditation Committee. The International Conformity Assessment Committee tracks international developments in Conformity Assessment and its activities are described in the section on Conformity Assessment.

The *National Policy Committee* (NPC) establishes operating rules for the Committee on Education. The NPC also develops policies for handling SDO input to ANSI for its approval. The Board of Standard Review (BSR) responds to the public comments on proposed ANSI standards produced by the notice in the ANSI weekly publication *Standards Action*. The Executive Standards

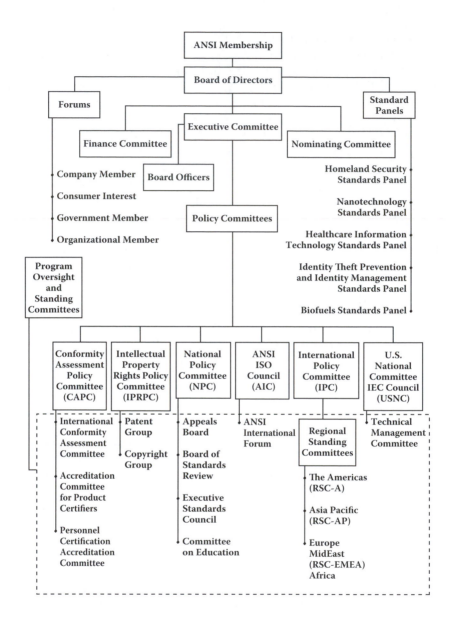

FIGURE 8.1
ANSI organization chart.

Council is also involved, as is the Appeals Board, which handles appeals of actions on proposed American National Standards.

The *ANSI ISO Council* (AIC) manages the ISO "national committee" activities of the ANSI International Forum, which includes the U.S. ISO Technical Committees and Subcommittees involved in developing ISO Standards.

The *International Policy Committee* (IPC) oversees the Regional Standing Committees (RSCs): RSC-A for the Americas, RSC-AP for Asia-Pacific, and RSC-EMEA for Europe, the Mideast, and Africa.

In the international and regional arenas, ANSI promotes the submission of its standards in those forums and also encourages the U.S. adoption of suitable International Standards where appropriate. ANSI also monitors and contributes to standards and conformity assessment activities in the *regional* organizations (Chapter 7) including the European Union (EU), the Pacific Area Standards Congress (PASC), the Pan American Standards Commission (COPANT) and the Pacific Accreditation Cooperation (PAC). In addition, ANSI is a member of the Inter-American Accreditation Board (IAAC) via the ANSI-ASQ American National Accreditation Board (ANAB).

The *U.S. National Committee of the IEC Council* (USNC) provides the interface with the IEC with the help of the Technical Management Committee. U.S. Technical Advisory Group (TAG) representatives are "certified" to the IEC for its Technical Committees and Subcommittees. Earlier in the twentieth century, the American Institute of Electrical Engineers (AIEE), one of two parent organizations of the IEEE, served as the U.S. National Committee of the IEC for a number of years. All of the organizational entities below the "policy level" (those in the dashed box in Figure 8.1) are monitored by the Program Oversight and Standing Committees as indicated.

ANSI President and Chief Executive Officer S. Joe Bhatia has provided an outline of how ANSI functions and what he expects in the future (ASTM 2007). Of particular interest, in regard to the new United States Standards Strategy (USSS), is his request: "I encourage everyone reading this article to go to the Web site (www.us-standards-strategy.org), read the Strategy, view the endorsement and implementation reports, and record what your organization is doing to support and assist in implementation."

The United States Standards Strategy (USSS)

The USSS was published in December 2005 and replaces the first National Standards Strategy (NSS), published in August 2000, because of substantial changes in the standards environment. The rationale for a Standards Strategy springs in substantial part from the diversity of the participants in the U.S. system and the need for a common understanding of standards issues. The USSS does not address conformity assessment but there is a link on the ANSI Web page (reference materials) that takes one to the second edition of *The National Conformity Assessment Principles for the United States*. The USSS recognizes the "new reality" of standards setting by *consortia* and the observance of World Trade Organization (WTO) principles for avoiding technical barriers to trade produced by standards.

The USSS is accompanied by "assigned tactics" for ANSI, Standards Developers, Government, Industry, Consumers, U.S. Stake Holders and Academia; the tactics are encapsulated briefly as follows:

1. Strengthen Participation by Government
2. Address the Environment, Health, and Safety
3. Improve Responsiveness to Consumer News
4. Promote Internationally Recognized Principles
5. Encourage Common Governmental Approaches
6. Prevent Standards from Becoming Trade Barriers
7. Strengthen International Outreach Programs
8. Improve Process for Development and Distribution
9. Promote Cooperation and Coherence within the System
10. Establish Standards Education as High Priority
11. Maintain Stable Funding Models
12. Standards Supporting National Priorities

In 1990, Karen A. Frenkel included some material about the (then-) perceived shortcomings of the ANSI arrangement with regard to the ability of the United States to meet the challenge of the Europe 1992 act. Frenkel's paper included a summary of a proposal by NIST's Stanley I. Warshaw, a prominent standards expert, to replace ANSI with a new body modeled after the Standards Council of Canada (see a brief description of the SCC in a later section). Many of the issues included in Warshaw's proposal have been addressed, primarily through the passage of the National Technology Transfer and Advancement Act (P.L. 104-113), called the NTTAA for brevity, and other actions taken by NIST and by ANSI. More details on the NTTAA appear in Chapter 11.

Henk J. de Vries (1999), a noted author on standards, has summarized the expected functions of a National Standards Organization. Comparing ANSI's functions outlined here with Vries' summary indicates that ANSI is a full-service one-stop organization that meets the needs of the U.S. Standards and Conformity Assessment community.

The National Conformity Assessment Principles for the United States

As a companion piece to the United States Standards Strategy, ANSI approved the second edition of the *National Conformity Assessment Principles for the United States* in May of 2007. It is based on and supplements the World Trade Organization's Agreement on Technical Barriers to Trade (TBT). The cornerstone of the *Principles* is the ISO/IEC 17000:2004 standard *Conformity Assessment—Vocabulary and general principles* described in detail in Chapter 2. The *Principles* are briefly outlined as follows:

I. Purpose: "The concise and clear presentation of these principles is intended to promote national and international understanding and recognition of competently conducted conformity assessment processes resulting in increased acceptance of U.S. products within national and international markets."

II. ISO/IEC 17000:2004 defines conformity assessment as "demonstration that specified requirements relating to a product, process, system, person or body are fulfilled. Conformity assessment includes sampling and testing, inspection, supplier's declaration of conformity, Certification, and management system assessment and registration. It also includes accreditation of the competence of those activities by a third party and recognition (usually by a government agency) of an accreditation program's capability."

See Chapter 5 and the first edition of ISO/IEC 17011, *Conformity Assessment—General requirements* for accreditation bodies accrediting conformity assessment bodies.

III. Applicability of principles: The *Principles* are broadly applicable to the varied activities of conformity assessment, aiming to achieve the goals of conformity assessment in a cost-effective manner.

IV. Definitions: The definitions of ISO/IEC 17000 are followed with few variations. Robert Hunter (2005) has compared the changes in the ISO/IEC 17000 definitions from the former definitions in ISO/IEC Guide 2.

V. Conformity assessment principles: Conformity assessments avoid the creation of obstacles to national and international trade and are based on international guides and standards. The processes are open, transparent, and driven by sector-based standards and practices. The goal is to meet both regulatory and market needs with a single process that is fairly applied to all CABs seeking *accreditation*.

VI. Suggested reading: Includes the WTO Agreement on Technical Barriers to Trade (TBT).

The National Voluntary Laboratory Accreditation Program (NVLAP)

As the storm clouds of Europe 1992 gathered in early 1990, the standards community began to realize the seriousness of the conformity assessment issues that Europe would raise for U.S. exporters. The issue was especially acute in the electromagnetic compatibility (EMC) testing field where European regulators would not accept test data from laboratories outside the EC. In casting about for solutions, this author wrote a memo in 1990 to the (then) chairman of the ESMC Committee of the Computer & Business Equipment Manufacturers Association (CBEMA, now INCITS) suggesting the use of the

NIST National Voluntary Laboratory Accreditation Program (NVLAP) to address that issue. The memo is excerpted below:

> There has been in existence for a decade a suitable body in the NIST for providing accreditation to Certifiers. The program, described in 15 CFR Subtitle A, Part 7, provides the statutory basis for the National Voluntary Laboratory Accreditation Program Procedures (copy enclosed). The suitability of this vehicle in the international certification arena would appear to follow from Para. 7.2(d) which reads:
>
>> NVLAP is carried out to be compatible with and recognized by domestic, *foreign*, and *international systems* for laboratory accreditation so as to enhance the universal acceptance of test data produced by NVLAP-accredited laboratories (emphasis added). A number of "merchant EMC testing laboratories" currently are accredited to enhance the international acceptance of their test data. Let's not reinvent the wheel and loose critical time in the process. The program is funded and in place. *Why not NVLAP?*

NVLAP, and a private sector accreditor, the American Association for Laboratory Accreditation (A2LA), didn't become the accreditors of choice for EMC labs for several more years. The NVLAP program clearly meets the ISO/IEC 17000 definition of an *accreditation body* in 3.2 and the Note at 3.2. More on the NVLAP appears in Chapter 11. NVLAP accreditation and the NVCASE program were the bases for the EMC and telecom CABs inclusion in the U.S.–EU Memorandum of Understanding mentioned in Chapter 6.

Standards, Conformity Assessment, and Trade

As pointed out in the National Research Council report *Standards, Conformity Assessment, and Trade* (NRC 1995), "Significant improvement is needed in the U.S. system for assessing conformity of products and processes to standards." This critique was based on the fact that conformity assessment in the United States was complex, fragmented, and redundant at all levels (federal, state, and local). The report urges the government to get out of conformity assessment activities and let the private sector do that job. The report also recognizes the need for government oversight of critical areas like public health and environmental concerns, implicitly endorsing the Nationally Recognized Testing Laboratory (NRTL) program of the Occupational Safety and Health Administration (OSHA) as an example of the proper role of government (see Chapter 2). The recommendations of the NRC 1995 report were embodied in legislation in the NTTAA Public Law referenced earlier. Government agencies are required by the NTTAA to use voluntary standards where possible and to task NIST with oversight of the program, including the improvement of conformity assessment at all levels. A perusal of the NIST Special Publication 903 (Hyer 2001) shows the diversity of the private sector conformity assessment bodies in the United States and the diversity in the *accreditation bodies* for those CABs. Most of the accredited CABs cited NVLAP, A2LA, or the Standards Council of Canada (SCC) as *accreditors*, some having multiple accreditations.

ANSI and ANAB Roles in Accreditation in the United States

As shown in Figure 6.1, ANSI has established Accreditation Services that accredit Product Certification Programs and Personnel Certification Programs. These programs provide confidence to the clients of those Certification programs that the Certifier's product (Certification) is acceptable in the marketplace. The ANSI Accreditation Services program is greatly enhanced by its recognition by the NIST National Voluntary Conformity Assessment System Evaluation (NVCASE) Program. More on NVCASE appears in Chapter 11.

The Product Certification Programs are accredited in accordance with ISO/IEC Guide 65—*General requirements for bodies operating certification systems* and other appropriate International Standards including ISO/IEC 17025—*General requirements for the competence of testing and calibration laboratories.*

ANSI also accredits Personnel Certification Programs in accordance with ANSI/ISO/IEC 17024—*Conformity assessment—General requirements for bodies operating certification of persons.* ANSI/ISO/IEC 17024 is of course the U.S. national adoption of the ISO/IEC International Standard.

ANSI joined with the American Society for Quality (ASQ) to form the (ANSI–ASQ) American National Accreditation Board (ANAB). This organization replaced an earlier arrangement that did not meet the requirements of ISO/IEC 17011. The ANAB has two accreditation programs:

Accreditation of Quality Management Systems (QMS) Registrars

Accreditation of Environmental Management Systems (EMS) Registrars

The Quality Management Systems Registrars evaluate the conformance of QMS systems in accordance with the International Standards in the ISO 9000 series of Quality Management and associated standards.

The Environmental Management Systems Registrars evaluate EMS organizations that are involved in the International Standards in the ISO 14001 series on Environmental Management Systems (a very substantial set of standards).

U.S. Accreditation Activity of the Standards Council of Canada (SCC)

The Standards Council of Canada is a federal Crown corporation whose mandate is to promote efficient and effective standardization. The SCC reports to the parliament via the Minister of Industry. The SCC is located in Ottawa and oversees Canada's National Standards System (www.scc.ca/en). The SCC, pursuant to the U.S.–Canada Free Trade Agreement, has been accrediting U.S. CABs for a number of years. Enabled by an Order in Council

(SI/2002-121), amending the Standards Council of Canada Act, SCC is now (as of 2002) authorized to accredit CABs in all WTO member countries! The SCC accredits four types of CABs: product certification bodies, testing and calibration laboratories, management systems registration bodies, and personnel certification bodies.

American Association for Laboratory Accreditation (A2LA)

The American Association for Laboratory Accreditation is a nonprofit, nongovernmental, public service membership society. A2LA provides services in training as well as laboratory accreditation. The laboratory accreditation is based on ISO/IEC 17025. A2LA is recognized widely via a number of bilateral and multilateral recognition agreements. A2LA is recognized in forty-six countries and thirty federal, state, and local agencies. A2LA accreditation of EMC Test Laboratories, for example, is recognized by the Federal Communications Commission (FCC). Many of the laboratories that advertise in the ASTM Standardization News are A2LA accredited.

Some Representative U.S. Bodies for Non-ICT Standards and Conformity Assessment

Limited space prevents the coverage of all but the most significant U.S. bodies.

American Society of Mechanical Engineers-International (ASME)

ASME was founded in 1880, primarily in response to the increasing number of steam boiler explosions. As an "engineering society," ASME develops standards and codes as well as conformity assessment schemes. ASME is best known for its Boiler and Pressure Vessel Code (BPVC), a multivolume collection of standards, but also has approximately 600 standards on "everything mechanical." In addition to safety codes, the ASME has developed Performance Test Codes to permit the evaluation of the efficiency of boilers and related equipment. ASME prepares standards for nuclear components and has a Certification mark that is recognized in more than 100 countries. All kinds of *fasteners* are subjects of ASME standards as well as escalators and elevators. ASME is active in the International Standards arenas and has working relationships with NIST and other government agencies such as the Department of Defense. More than 100 ASME standards are referenced in government regulations. ASME has substantial outreach and educational programs, and it publishes two research periodicals, the *Journal of Heat Transfer* and the *Journal of Applied Mechanics*, as well as the membership publication *Mechanical Engineering*. ASME has approximately 120,000 members. Its Web site address is www.asme.org.

American Society for Testing and Materials International (ASTMI)

The ASTMI is a nonprofit organization formed in 1898 and originally named the International Association for Testing Materials (IATM). The impetus for the formation of ASTM, as it was later known, was the urgent need of the industrial age, including especially the Pennsylvania Railroad, for steel and other building materials that met the needs of the *users*. The Pennsylvania Railroad in that day found that informal "craft methods" of producing steel of the required properties was unsatisfactory. Although the bulk steel manufacturers resisted early standards, the ASTM became the forum for the development of *consensus* standards, standards that met the needs of both producer and user. In 2001, ASTM changed its name to include the "International," reflecting the worldwide status of its international standards portfolio of some 12,000 documents.

ASTMI is located in West Conshohocken, Pennsylvania, just northwest of Philadelphia. ASTMI is a scientific as well as a membership organization with outstanding expertise in analytical chemistry. Its standards go through an extensive consensus process, including open committee participation and voting by all members. Major topics covered by ASTMI standards include metals, paints, plastics, textiles, petroleum, construction, energy, the environment, consumer products, medical services and devices, electronics, and many other areas of expanding competence. ASTMI's development of the Light Sports Aircraft for the Federal Aviation Administration (FAA) was mentioned in Chapter 4. Various government agencies have adopted thousands of ASTMI standards, thus converting them from (voluntary) standards to regulations. ASTMI does not accredit laboratories but publishes a large guide to accredited laboratories. In addition to its standards, ASTMI published special reports as well as its membership publication *Standardization News (SN)*. SN tracks external standards developments as well as internal ones. ASTMI has a robust program of outreach to other countries, including Memorandums of Understanding with numerous national and regional standards organizations. More information appears on its Web page at www.astm.org.

Electronics Industries Alliance (EIA)

The EIA, formerly known as the Electronics Industries Association, is a trade association that started life as the Radio Manufacturers Association and later as the Radio–Electronics–Television Manufacturers Association (hence the famous RETMA Rack). Most EIA standards are submitted to ANSI for recognition, some have replaced MIL (Military) Standards. The EIA is located in Arlington, Virginia, and is active in several Technical Committees and Subcommittees of the IEC. EIA is home for the JEDEC Solid State Technology

Association, formerly the Joint Electron Device Engineering Council (JEDEC). The EIA is also the National Standards Organization (NSO) for the IECQ Scheme operation in the United States.

In addition to the JEDEC organization, the EIA is an umbrella organization for the following major bodies:

Telecommunications Industry Association (TIA), which, among other things, develops ICT standards (more on ICT bodies in an upcoming section)

Government Electronics and Information Technology Association (GEIA)

Electronic Components, Assemblies and Materials Association (ECA) and several affiliate bodies

Through its TIA Sector body and several affiliates, EIA is also involved in the ICT area.

National Electrical Manufacturers Association (NEMA)

NEMA is a trade association that develops standards for many kinds of electrical equipment. It was founded in 1926 and is located in Washington DC. Some of its standards are recognized by ANSI and by the government. NEMA champions the use of the National Electrical Code (NEC/NFPA) and the National Electrical Safety Code (NESC/IEEE) to promote electrical safety indoors and outdoors, respectively. NEMA is very active in support of ANSI and the U.S. National Committee (USNC) of the IEC. In addition to the IEEE and NFPA, NEMA works closely with the ASTM and Underwriters Laboratories (UL). NEMA participates in the IECEE (Chapter 6) and the ISO. NEMA's "wall plug" specifications are probably among its best-known standards.

National Fire Protection Association (NFPA)

The NFPA is an independent nonprofit organization that was founded in 1896. It is located in Quincy, Massachusetts, and has more than 50,000 individual members and many organizational members. It produces and maintains nearly 300 National Fire Codes and many of its standards are recognized by ANSI. Every firefighter is schooled in NFPA codes for the prevention and suppression of fires. NFPA investigates major fires and develops national statistics for fire losses. Sprinkler protection systems to minimize fire damage are mandated by many state and city codes, but older buildings are frequently exempted from compliance. The cost of installing sprinkler systems in old buildings is small compared to the cost of property damage and lives lost due to major fires. One of NFPA's best-known standards is ANSI/NFPA 70, the *National Electrical Code (NEC)*. The various parts of the NEC are developed by Code Making Panels and are widely reviewed by the public. Periodic revisions of the NEC are approved by the National Conventions. The NEC, a voluntary standard, is a good example of a "Pre-Reg" type of standard mentioned in Chapter 3 since it known in advance that it will be adopted widely in most U.S. states, counties, and cities, sometimes in modified form. NFPA standards actions can be appealed to the NFPA Board of Directors.

Underwriters Laboratories (UL)

Underwriters Laboratories is an independent nonprofit standards development and conformity assessment body that is based in Northbrook, Illinois. UL was founded in 1894 in response to problems with electrical fires resulting from early electrical fixtures, which had (flammable) wood parts. UL has nearly 1,000 standards of which about half are recognized by ANSI. Most UL standards are safety oriented, but UL is also a registrar for Quality Management Systems (QMS). The QMS registration is sometimes used to supplement conformity assessment programs in a manner roughly analogous to the way some EU directives call for QMS (ISO 9000) standards to supplement type approval and surveillance activities. Many government agencies reference UL standards for safety, including the U.S. Coast Guard for which UL tests life preservers for regulatory compliance. There is a "theory" of standards that says that the development of the standards and the conformity assessment activities should be done by separate organizations. UL has managed to accommodate both activities in the same organization. For many years UL had a virtual monopoly on safety testing and was the only agency recognized in the NEC. UL now has competitors, some for-profit, that are Nationally Recognized Testing Laboratories (NRTLs). Most of UL's competitors in product-safety testing use UL standards.

UL develops and maintains its standards by using technical committees called standards panels. The process is open and accompanied by full public participation. Each UL requirement is supported by a rationale to show its legitimate motivation. UL pioneered the adoption of IEC standards for safety in the United States that made it unnecessary for U.S. producers of certain products to meet different standards for export. The mode of operation of UL was outlined in Chapter 2 on conformity assessment. The surveillance task in UL is performed by the "Follow-Up Service" that makes unannounced inspections at the production sites of UL certified equipment and components. UL is a port of entry to the IEC CB Scheme described in Chapter 6. UL also works with the EIA as the National Supervising Inspectorate (NSI) of the IECQ system. UL has laboratories and offices worldwide that provide services to both U.S. and foreign clients. Several UL standards became binational (with Canada) and trinational including Mexico. UL's family of companies and service providers include 58 laboratories and 127 inspection centers around the world. UL has promoted the concept of "transportable conformity assessment" (Mader and Haataja 1999).

Some Representative U.S. Bodies for ICT Standards and Conformity Assessment

Again, space limitations prevent the coverage of all but a few representative U.S. bodies.

Alliance for Telecommunications Industry Solutions (ATIS)

The demise of the Bell System in 1983 left a vacuum in the telecom standards area. Many Bell System standards were de facto and some not even published in open literature. The Exchange Carriers Standards Association (ECSA) and others got together to form a standards organization called T1 (Lifchus 1985). T1 was a regional ICT body. Although located primarily in the United States, T1 became an ANSI accredited standards developer and proceeded to provide the standards needed by the evolving ICT infrastructure. After twenty years, it became apparent that a new umbrella organization was needed to cope with the rapidly developing ICT areas nationally as well as regionally and internationally. In 2004, that new organization became ATIS.

ATIS is huge indeed. It has more than 1,100 industry professionals from more than 350 companies that aim to develop standards using flexible and open approaches. ATIS is located in Washington DC and interfaces with many government agencies and public and private ICT organizations. ATIS procedures are multimodal; in some cases it adopts standards (like IEEE and ITU standards), in other cases it maintains existing T1 standards. Some of the old T1 committees have been retained and rebadged; others have been terminated. ATIS has more than twenty-three industry committees. In still other cases ATIS develops totally new forward-looking standards to assist industry in developing new equipment and services. ATIS teams with TIA to support the Administrative Council for Terminal Attachments (ACTA). ACTA took over the FC Part 68 management of this area (see the modular plug/jack discussion in Chapter 3). ATIS has adopted consortium models called Incubators for the rapid development of standards. ATIS, like most of its major players, are ANSI Accredited Standards Developers; thus ATIS leverages the ANSI accreditation to manage the large group. ATIS is a member of the ITU and ETSI; it was a founding member of the Global Standards Collaboration (GSC) that was prominent in the rapid development of International ICT Standards (Habara 1994). Achieving and maintaining interoperability is a primary goal of ATIS. The URL for ATIA is www.atis.org.

Institute of Electrical and Electronics Engineers (IEEE)

The American Institute of Electrical Engineers (AIEE) was formed in 1884, the same year as the Electrical Exposition at Philadelphia that was sponsored by the Franklin Institute (Freitag 1986). The Institute of Radio Engineers (IRE) was formed in 1912, the genesis of electronics engineering. The two organizations merged in 1963 to form the IEEE. IEEE is a very broad-based organization, the largest technical organization in the world, with members all over the world (a concept called "transnational"). Members are not only engineers but are physicists and scientists from many other related disciplines as well. There are several dozen societies and other entities in the IEEE that specialize in various sciences and technologies. Many of the AIEE members became members of the Power Engineering Society (PES) and the Industry Applications Society (IAS). Many IRE members became members of the Communications Society (Com-Soc). Still others became members of the emerging Computer Society (CS), the largest of the IEEE societies.

The inventors of the Internet protocol (TCP/IP) published their paper in the IEEE Transactions on Communications in 1974 (Cerf and Kahn 1974). The basic standard for the LAN/MAN (local area network/metropolitan area network) was prepared by the Computer Society. IEEE Std 802.3 laid the foundation for the family of IEEE 802® standards on which access to the Internet by wire and wireless now depends (Graube 1982). This family of standards is concerned mostly with the *physical* and *data link* layers of the ISO Open Systems Interconnection (OSI) model. Many of the 802 family of standards have been transposed to ISO/IEC standards by JTC 1. The IEEE enjoys special access to the International Standards community, including the IEC and the ITU. IEEE is accredited by ANSI and has a long history of working with ANSI. The IEEE standards organization, now called the IEEE-Standards Association (IEEE-SA), has pursued the consensus approach to highly technical standards. As outlined in Chapter 4, the IEEE-SA has also embraced a version of the Consortium Method of standards that makes use of the IEEE-SA "infrastructure" in providing Secretariat type functions (Gorman and Mills 2006). Technical education is one of the major thrusts of the IEEE whose various entities are well suited to address the multidisciplinary issues in ICT standards setting. IEEE's Web site is www.ieee.org.

InterNational Committee for Information Technology Standards (INCITS)

INCITS is a forum for information technology developers and users for formal IT Standards. INCITS is an ANSI-accredited SDO and is supported by the Information Technology Industry Council (ITI), a trade association of U.S. providers of IT products and services. INCITS was started as ANSI Accredited Standards Committee X3, a committee of the Computer and Business Equipment Manufacturers Association (CBEMA). INCITS is located in Washington DC, and has 1,700 members from 13 countries so it is a regional organization, like ECMA (originally called the European Computer Manufacturers Association), which was a sister organization.

The X3 standards provided the basic standards of the information age such as the ANSI/INCITS X3.4-1986 (R1997): Information Processing-Coded

Character Sets-7-bit American National Standard Code for Information Interchange, the famous ASCII character set. X3 was a forum for producers and users to exchange information on IT matters of interest. INCITS serves as the Technical Advisory Group (TAG) to ANSI for the Joint Technical Committee 1 (JTC 1) process, which has replaced X3. INCITS Technical Committees address standards issues in IT hardware, programming languages, biometrics, and radio frequency identification (RFID), and liaises with the IEEE. The INCITS Web address is www.incits.org.

Telecommunications Industry Association (TIA)

The TIA was formed as the telecommunication sector organization of the EIA in 1988 through a merger of the U.S. Telephone Suppliers Association and an existing EIA group. TIA standards are included in the EIA Standards and Engineering Publications catalog. TIA, through its GLOBALCOMM™ forums, is a leading trade organization for the ICT industry. TIA standards are basic to the cell phone industry. TIA is an accredited ANSI standards developer in a variety of technologies. TIA's Engineering Committees include 70 subcommittees and working groups with more than 1,500 individuals from 20 countries, making it a regional body. TIA is active in the Third Generation Partnership Project 2 (3GPP2) and Technical Advisory Groups for ISO, IEC, and JTC 1. TIA is active in CITEL (Chapter 7) as well as the ITU (Chapter 6). TIA is active in standards being developed by local, state, and federal governments, and the efforts to develop a suite of standards for the 4.9 GHz band. The TIA Project MESA, in cooperation with ETSI (Chapter 7) aims to facilitate next generation wireless communications for public safety. TIA is active in fiber optics, user premises equipment, and wireless communications and satellite communications. TIA members must be *firms*, not individuals. The Web address for TIA is www.tiaonline.org.

Standards, Conformity Assessment, and Accreditation in Canada

As explained in the following, there has been a long history of beneficial interaction between Canadian and U.S. bodies.

Standards Council of Canada (SCC)

As mentioned earlier, the Standards Council of Canada (SCC) is a Crown corporation with direct connection to the federal government in Ottawa. Thus, the SCC has oversight for all voluntary standards activities in Canada. The Standards Council performs the main coordination of standards activities in Canada and is the "National Committee" for Canada in the ISO and the IEC. The SCC has held leadership positions in the ISO committees developing the ISO 9000 and ISO 14000 families of standards. The SCC developed a national standards strategy at an early date. The functions and outreach of the SCC map pretty well with those of ANSI, outlined earlier. The Standards Council is an advisor to various government entities, including federal, provincial, and territorial governments. The SCC works closely with the federal government to provide standards-based regulations through its Intergovernmental Affairs and Trade (IGAT) Branch. Canadian standards can be included in laws and regulations by reference as well as by direct adoption. Canada has been a leader in the adoption of International Standards. The scope of the SCC includes the enhancement of Canadian competitiveness and the promotion of Canada's social and economic well being. The SCC has been involved in the WTO, NAFTA, and the Free Trade Area of the Americas (FTAA). Because Canada has two official languages, all standards and regulations must be published in English and French. Like ANSI, SCC accredits Registrars for Quality and Environmental Management Systems. Although electrical safety is the responsibility of the provinces and territories of Canada, it is stipulated in the Canadian Electrical Code that certification by an SCC accredited CAB is acceptable across Canada. More information on the SCC is available at www.scc.ca/en/.

Canadian Standards Association (CSA)

There are fewer standards and conformity assessment bodies in Canada than in the United States. The Canadian Standards Association (CSA), based in Rexdale, Ontario (near Toronto), is the most prominent standards organization in Canada. "CSA's mission is to understand and fulfill Canadian society's national and international needs for consensus standards and conformity assessment related to safety, quality and performance" (Browne 1995). CSA is accredited by the SCC. The Canadian Electrical Code (CEC), published in two parts, is prepared and published by CSA rather than by a separate body like the NFPA in the United States. Part I of the CEC borrows from the U.S. National Electrical Code and there are many similarities in the two codes. Part II of the CEC is a collection of standards covering most electrical items such as attachment plugs (rather than a NEMA standard like the United States). CSA and UL have cooperated on electrical product safety standards for many years. The CSA Standards on Electromagnetic Compatibility, for example, serves the function of the Federal Communications Commission (FCC) regulations in the United States. CSA works closely with the British Standards Institution (BSI) and partners with several National Standards Bodies.

CSA, like its counterpart UL, produces standards and is also a conformity assessment body (CAB). CSA has field offices for conformity assessment in most of the provinces and in other parts of the industrialized world, including the Far East. The Canadian Standards Association is an OSHA Nationally Recognized Testing Laboratory (NRTL) for conformity assessments in the United States and is accredited by ANSI. Thus a Certification client in either country has a choice, among others, of UL or CSA. CSA America, located in Cleveland, Ohio, has traditionally done testing for gas (including hydrogen) safety. The Cleveland operation is being expanded and its scope enlarged.

Standards and Conformity Assessment in Japan

Japan is selected here as an interesting case study of a prominent national system for standards and conformity assessment.

Ministry of Economy, Trade and Industry (METI) and the Japanese Industrial Standards Committee (JISC)

The Ministry of Economy, Trade and Industry (METI) is the reorganized and renamed Ministry of International Trade and Industry (MITI). METI, as a government agency, funds much of the standards and conformity assessment activity in Japan. The close federal relationship is similar to that in Canada (SCC) and in the United Kingdom (the British Standards Institution [BSI]). Although there are other standards developers in Japan, the JISC prepares most of the voluntary Japanese Industrial Standards (JIS). Japanese Industrial Standards development started in the post-WWII era. Draft standards are developed by a variety of sources including trade associations and scientific societies. The negotiation of the standards then takes place in the JISC. JISC then passes the completed standard on to the Agency for Industrial Science and Technology (AIST). AIST approves and announces the standard in the Official Gazette. The standard then goes to the Japanese Standards Association for publication. Standards developers operating outside the government-approved JIS system usually work through the AIST. The JISC is the National Standards Body (NSB) member of the ISO and IEC; the robust Secretariat support that it receives from the METI is partly responsible for its success. The JIS Certification mark on products is prized by Japanese consumers; the conformity assessment process is similar to that described in Chapter 2. ISO 9000 standards are very popular in Japan, which has an obsession with quality.

The government sponsorship of standards activities is an example of "industrial policy," also known as mercantilism by political economists (OTA 1992). Industrial policy has long been a bad word in the United States, but it has been largely responsible for the "Japanese Miracle." The Japanese use of standards and conformity assessment as protectionist trade barriers has received a lot of attention (Krislov 1997). These policies have produced considerable trade friction between the two major trading partners (United States and Japan). The OTA report mentions the famous Japanese rejection of American-made baseball bats, a trade dispute that was eventually settled at the GATT (OTA 1992, 84, ref. 52]. Other examples are mentioned on page 147 of Krislov (1997). On the other hand, Japan does an excellent job of promoting and supporting standards that encourage the growth industries pursuant to industrial policies (McIntyre 1997). Japan plays a sophisticated standards game.

The outreach of Japan's standards community is substantial, much of it handled by the Japanese Standards Association programs in education. Japan has made substantial (multibillion dollar) investments in the Asia Pacific Telecommunity in the hope that regional ICT standards will reflect Japanese standards. As of 1993, Japan had made large grants and $4 billion in infrastructure loans in Asia (McIntyre 1997, 86). In September of 2001, JISC released its *Standardization Strategy* with main points as follows:

1. Responding to market and social needs
2. Strategy for international standardization
3. Integration of research and development and standardization

One point called for strengthening cooperation with the Pacific Area Standards Congress (PASC) and the Asia Pacific Economic Cooperation (APEC). Another point is that "Japan also understands that in some areas, SDOs other than ISO/IEC are developing international prevailing standards."

Japan's use of standards as industrial policy have supported a technical infrastructure that is far ahead of many competitors. Standards work has the attention of top managers in Japan, whereas standards work in the United States is frequently assigned to lower-level managers.

Standards and Conformity Assessment in Germany

In the following, the focus is on the German National Body DIN. DIN is similar to ANSI and the DKE is analogous to the U.S. National Committee of the IEC.

Deutsches Institut fur Normung (DIN)

In English, the German Institute for Standardization DIN is the National Standards Body (NSB) of Germany. It has been centered in Berlin since 1917 and has facilities all over Germany and the world. DIN is a nonprofit organization and very large compared to similar organizations. While many Americans are unfamiliar with ANSI, all Germans are familiar with DIN because its standards and activities are so pervasive. In exchange for its official recognition by the German government as Germany's National Standards Body, DIN is committed to serving the public good (OTA 1992, 62). DIN operates by consensus but is more centralized than the ANSI membership and includes producers, academics, users, government, and trade unions. DIN standards are well respected in Europe and the world. Once a standard is agreed on, it is reviewed by the Standards Inspection Office in DIN to assure that it does not conflict with existing standards. The principles of DIN standards development are similar to those of ANSI's Essential Requirements but are focused more on trade and economic factors. DIN responds to requests from the federal government for the preparation of standards to support regulations. The use of DIN standards is viewed as a means of deregulation in Germany. The public–private partnership of the government and private sector (like DIN) has been a long-time practice in Germany and the formal documentation of that arrangement, the "Standards Agreement," was a model for the "New Approach" in the EU. DIN's activities are supported in part by DIN Group, a profit-making subsidiary.

German business support of DIN is high. "Moreover, businesses have a strong incentive to adopt DIN standards, since the use of DIN standards shifts liability to the user, and insurance companies often refuse to grant policies to those whose products fail to carry the DIN testing and inspection marks" (OTA 1992, 64). The director of DIN, Torsten Bahke, was interviewed by ASTM's *Standardization News* in 1993. Bahke emphasized the fact that German firms know that the transaction costs of converting from one standard to another is five to seven percent of the total contract. He also stated, "We have come to a point where standardization bodies do not compete with each other. We standards developers have to serve our stakeholders" (ASTM 2003, p. 31). DIN is very sensitive to the cost of standardization as evidenced in its report on the economic benefits of standardization (DIN 2001). That report will be reviewed in Chapter 12. DIN is very active in International and Regional standards bodies and holds many key positions. The German National Committee of the IEC is the Deutsche Elektrotechnische Kommission (DKE) in DIN and VDE (Verband Deutscher Elektrotechniker). VDE, the German Association of Electrical Engineers, performs many of the functions of the IEEE and UL. The reduction of the importance of strictly national standards is well illustrated in Figure 3.1 of the OTA report (OTA 1992, 63). Over a six-year period between 1984 and 1990, DIN's work on purely national standards fell by half (in the run-up to the Europe 1992). During the same period, DIN's activities in European standardization roughly tripled.

DIN has entered cooperation agreements with the United States, Russia, China, and India. These agreements and other DIN actions are designed to eliminate trade barriers. DIN has a special committee for research and development phase standardization, a strategy also being used by Japan. In the march to higher-level metrication (more on that in Chapter 12), the DIN connectors and DIN rails are becoming more widely known in the United States. The DIN Web page is www.din.de.

References

American Society for Testing and Materials (ASTM), 2003, Interview with Torsten Bahke, Director, Deutsches Institut fur Normung, *ASTM Standardization News* 31(9): 26–31.

American Society for Testing and Materials (ASTM), 2007, An Interview with American National Standards Institute President and CEO S. Joe Bhatia, *ASTM Standardization News* 35(2), 26–29.

Browne, Bernice, 1995, CSA Consumer Program Taps Grassroots Concerns, *ASTM Standardization News* 23(3): 38–41.

Cerf, Vinton G., and Robert E. Kahn, 1974, Protocol for Packet Network Intercommunication, *IEEE Transactions on Communication* 22(5): 637–648.

Deppa, Shelley W., 1997, The Unchallenged Assumption of U.S. and International Standards Harmonization, *SES Standards Engineering* 49(2): 1–5.

de Vries, Henk J., 1999, *Standardization: A Business Approach to the Role of National Standardization Organizations*, Amsterdam, Kluwer Academic Publishers, pp. 91–134.

Deutsches Institut fur Normung (DIN), 2001, Economic benefits of standardization— Summary of Results; Final Report and Practical Examples.

Freitag, Harlow, ed., 1986, *Electrical Engineering: The Second Century Begins,* New York, IEEE Press, 1986, pp. 4–5.

Frenkel, Karen A., 1990, The Politics of Standards and the EC, *Communications of the ACM* 33(7): 41–51.

Gorman, Judy, and Steve Mills, 2006, The Evolving Spectrum of Standards Development Processes, *SES Standards Engineering* 58(2): 1–6.

Graube, Maris, 1982, Local Area Nets: A Pair of Standards, *IEEE Spectrum* 19(6): 60–64.

Habara, Kohei, 1994, Cooperation in Standardization, *IEEE Communications Magazine* 32(1): 83.

Hunter, Robert D., 2005, The New ISO/IEC 17000, *SES Standards Engineering* 57 (4): 12–13.

Hyer, Charles W., ed., 2001, *Directory of U.S. Private Sector Product Certification Programs*, NIST Special Publication 903.

Krislov, Samuel, 1997, *How Nations Chose Product Standards and Standards Change Nations,* Pittsburgh, PA, University of Pittsburgh Press, p. 147.

Lifchus, Ian M., 1985, Standards Committee T1—Telecommunications, *IEEE Communications Magazine* 23(1): 34–37.

Mader, Donald A., and David C. Haataja, 1999, UL's Free Trade Policy: Transportable Conformity Assessment, *SES Standards Engineering* 51(2): 1, 16.

McIntyre, John R., 1997, *Japan's Technical Standards: Implications for Global Trade and Competitiveness,* Westport, CT, Quorum Books, p. 80.

National Research Council (NRC), 1995, *Standards, Conformity Assessment, and Trade: Into the 21st Century*, Washington DC, National Academy Press, p. 154.

Office of Technology Assessment (OTA), 1992, *Global Standards: Building Blocks for the Future*, Washington DC, U.S. Congress, Office of Technology Assessment, p. 84.

Oksala, Stephen, 1996, National versus International Standards: Products and Processes, *SES Standards Engineering* 48(6): 1–7.

Toth, Robert B., ed., 1991, *Standards Activities of Organizations in the United States*, NIST Special Publication 806.

9

Provincial (State) and Local Standards and Conformity Assessment

Introduction

Just as nations need some unique standards, despite the existence of International Standards, states and provinces may also need "unique" Provincial Standards and conformity assessment mechanisms despite the existence of National Standards and conformity assessment systems. The *general* principle is that state (provincial) standards may be more stringent than national standards, but not less stringent. That principle was challenged in 2008 by the Environmental Protection Agency's refusal to permit California and other states to reduce greenhouse-gas emissions from automobiles. In most other situations, states can, as standards users, select standards from a menu of international, regional, and national standards. States may also require a standard due to "local" conditions for which no National Standard exists (with the possible exceptions like that indicated above). Because of the political inertia of the federal government, state governments can frequently act more quickly that the federal government. Examples of unique state standards due to varying climate and other conditions include:

California has stricter air pollution standards because of its air quality problems and also requires building codes that emphasize earthquake resistance, standards not required in all states.

Florida has strict building codes that emphasize hurricane resistance, standards not required in North Dakota, for example.

Chicago, after the great fire of 1871, developed more stringent fire-resistant building codes to avoid a repetition of the tragedy.

Several states, including California, outlawed the gasoline additive MTBE (methyl tertiary butyl ether) before a National Regulation became available.

The National Electrical Code (NEC)

The National Electrical Code, NFPA 70, is an American National Standards Institute (ANSI)-recognized American National Standard. The NEC is revised every three years and is open to public comment. The NEC is a prime example of a "pre-reg" standard. While technically a "voluntary standard," it is known in advance that this *model code* will be incorporated in most state and local electrical codes. Typically, the NEC is adopted without significant change in twenty-five states and adopted with modifications in nineteen states. Virtually all electrical safety standards, such as UL standards, use the NEC as a foundation. As mentioned in Chapter 8, the NEC was used in part by Canada in the development of its Canadian Electrical Code. Various parts of the NEC are being developed as trinational (Canada, United States, and Mexico) electrical standards pursuant to the North American Free Trade Agreement mentioned in Chapter 7. In many European countries, IEC 364 is used as the basis for a "national electrical code."

The City of Austin, Texas, Adoption of the National Electrical Code (NEC)

Most major cities have an Electrical Code. As an example of the way the NEC is incorporated by reference in the Code of the City of Austin, Technical Codes, Article 4, Electrical Code, has four sections:

Section 25-12-111 Electrical Code

Section 25-12-112 Citations to the Electrical Code

Section 25-12-113 Local amendments to the Electrical Code—Administration

Section 25-12-114 Local amendments to the Electrical Code—Technical

Section 25-12-111 reads in pertinent part as follows:

(A) The National Electrical Code, 2005 edition, published by the National Fire Protection Association (2005 Electrical Code) is adopted and incorporated into this section, including Annex G, with deletions and amendments in Subsections (B) of this section and Sections 25-12-113 and 25-12-114.

(B) The following provisions of the 2005 Electrical Code are deleted: [There follows a list of 39 NEC sections and Table 310.5.]

(C) The City Clerk shall file a copy of the 2005 Electrical Code with the official ordinances of the City.

Model Building Codes Adopted by States

Early in the twentieth century, a trend to replace unique local building codes by model building codes was established. The Building Officials Code Administrators (BOCA) developed a regional model building code that was used in the eastern and central part of the United States. Another model code was the International Conference of Building Officials (ICBO) that was adopted by states in the western United States. The third major regional building code that emerged was the Southern Building Code Congress International (SBCCI) that was adopted by a swath of states in the southern region of the country. These building codes were modular, including, for example electrical codes, plumbing codes, gas codes, and fire prevention codes. The result was that the nation was covered by a "crazy-quilt" adoption of the three codes and various combinations of them as well as several truly unique state codes. Architects and engineers had to learn the different codes, especially if doing projects across state lines, at a substantially higher transaction cost than would result from a common set of model codes.

In the 1990s it was realized how inefficient having several model building codes was. A sea change occurred when the ICBO agreed to stop development of the several model codes and establish the International Code Council (ICC) in 1994. Roy Fewell (1997) describes the ICC process for developing model codes. Thus a variety reduced set of International Codes was developed by the year 2000. The portfolio of modular-coordinated International Codes includes the following (from the ICC Web page: www.iccsafe.org):

- International Building Code (IBC)
- International Energy Conservation Code (IECC)
- International Fire Code (IFC)
- International Fuel Gas Code (IFGC)
- International Mechanical Code (IMC)
- International Plumbing Code (IPC)
- International Private Sewage Disposal Code (IPSDC)
- International Property Maintenance Code (IPMC)
- International Residential Code (IRC)
- International Zoning Code (IZC)
- ICC Electrical Code Administrative Provisions (ICCEC)

The International Building Code (IBC) is the cornerstone of the ICC family of codes; the IBC itself takes over (adopts or references) many other standards such as ASTM standards (Johnson 2003). The IBC-referenced standards include requirements for materials, installation, testing, and design. As often happens in adopted standards, the normative documents themselves contain references that also become part of the code.

Forty-seven states and many local jurisdictions have adopted one of more of the ICC family of model codes as is documented in the "Government Relations" link on the ICC Web page. Texas, for example, has adopted the 2003 IBC effective statewide. California has approved the adoption of the 2006 IBC and IFC to become effective in 2008. The National Institute of Standards and Technology (NIST) did a three-year study on the hazards of tall buildings as a result of the collapse of the World Trade Center Twin Towers on September 11, 2001. The results of the study are being incorporated in the 2007 Supplement of the IBC (NIST 2007). More information is on the NIST Web page at http://wtc.nist.gov.

The New York State Experience

Historically, the Building Codes in New York State (the New York State Uniform Fire Prevention and Building Code) did not follow the regional building codes of its neighboring states (Thompson 2002). As a result, construction and maintenance expenses for buildings in the state were significantly greater than those in other states. Moreover, many of the old requirements in the New York Building Code included *descriptive provisions* (prescriptive or design based) that did not make use of the superior *performance provisions* that improved the safety, economy, energy conservation, and other requirements that are available in modern technologies. This is an example where a state unique code doesn't make sense. To solve those problems, a large collaborative effort was made to produce a new New York Building Code based on the International Building Code. That effort involved users as well as builders and others, and included the New York Energy Conservation Construction Code (ECCC) and other modern developments. In addition to reducing construction costs, the new building code thus reduced the cost of ownership.

Texas Commission on Fire Protection Adoption of Certain NFPA Standards

Section 531.7, *Adopted Standards,* are part of the Texas Fire Alarm Rules (17 TexReg 8504; Texas Commission on Fire Protection). Section 531.7(a) reads in pertinent part (a legacy example):

> (a) The Commission [Board] adopts by reference those sections of [in their entirety] the following copyrighted **minimum** standards [and] recommendations, **and appendices concerning fire alarm, fire detection, or supervisory services or systems**, except to the extent they are at variance to sections of the **chapter**, [subchapter] the **Texas Insurance Code, Article 5.43-2**, or other state statutes. There follows a list of 24 NFPA

standards and codes (including the NEC) that are applicable to the Fire Alarm Rules.

This example is typical of the way NFPA standards (and codes) are "incorporated by reference" in a state standard.

State Highway Standards

The Interstate Highway System is, of course, covered by a special set of federal standards. This was necessary to establish uniformity in the Interstate Highway System. The federal standard for an interstate highway includes the following requirements (McNichol 2003):

At least two lanes in each direction

Each lane is 12 feet wide

The far left side has a 4-foot wide shoulder

The right side has a 10-foot wide breakdown lane

The federal specifications clearly cannot be used for smaller state roadways. As a result, each state has its own set of standards for highway construction, although some of the pavement technology and other items later found their way into modern local roadways. The proof of concept for the interstate highway standard was established on a test road seven miles long built in Illinois. The whole program demonstrated how the federal government, the states, and private concerns could work together. The enabler of this cooperation was the formation, in 1914, of the American Association of State Highway Officials (AASHO). AASHO was a semiprivate forum where highway engineers could gather to work out issues with minimal political interference. The name was changed in 1973 to the American Association of State Highway and Transportation Officials (AASHTO). AASHTO has a very large portfolio of more than 1100 standards, some of which have been referenced in Federal Highway Administration regulations. AASHTO also provides a forum for many private sector SDOs that produce standards related to pavement (Kandhal 2003).

Texas and Missouri have traditionally been leaders in highway construction. The Texas Department of Transportation (TxDOT) specifications and standards are often used by highway departments in other states (James Mercier, personal communication 2007). TxDOT pioneered the use of the breakaway lamp pole to minimize damage and injuries when vehicles collide with them. The Texas Department of Transportation Standard Specifications for Construction and Maintenance of Highways, Streets and Bridges (June

1, 2004) runs more than 1,100 pages. It references standards developed by AASHTO and ASTM, as well as Department Material Specifications. The Standard Specifications rely on 40 CFR 261 to define hazardous materials or waste. Several conformity assessment provisions in the Standard Specifications are used to control the quality of the materials used and quality of the work done.

The MTBE Fiasco: State Standards Leading Federal Standards

The Federal Clean Air Act Amendment of 1990 required gasoline to have an oxygenate added to reduce air pollution in auto exhaust. The Environmental Protection Agency (EPA) did not specify the oxygenate to be added but left it up to refiners to meet a performance standard (seemed like a good idea at the time). The EPA permitted the use of MTBE in gasoline even though it reportedly knew about its potential threats to health. Refiners had previously used smaller amounts of MTBE in gasoline to raise its octane level after the use of lead for that purpose was banned. MTBE is highly soluble in water and slow to degrade, so a relatively small amount can contaminate a lot of water. The problem was made worse by the leakage of gasoline from filling stations and other storage tanks that polluted groundwater. Small amounts of MTBE had thus been tolerated for some time, but the higher levels subsequently used contaminated the drinking water in Santa Monica, California, so badly that it had to truck in drinking water! According to a *60 Minutes* story that aired on January 16, 2000, EPA officials stated "they were working to develop a standard."

Other areas suffered similar problems. Another case was the MTBE pollution of groundwater in east Austin, Texas, by leakage and spillage from a "tank farm" used by several oil companies to store gasoline in the Austin area. Much of the MTBE contamination came from the leaking tanks of gasoline retailers (Scheibal and Haurwitz 2001). The litigation surrounding that situation produced an expert witness who said "MTBE must not be dangerous since there is no standard for it" (Otto Friedrich Jr., personal communication 2000). The litigation was settled in 1996 (*U.S. Water News Online*, April 1996). Some technical details of the issue and the status of pending MTBE legislation appear in an ASTM article (Herman and Clark 2002). Banning the use of MTBE has encouraged the use of ethanol and other means to reduce air pollution.

The MTBE issue is an example of states acting to remedy a serious problem because of *federal inaction*. The MTBE fiasco might have been avoided or mitigated if the Office of Technology Assessment had not been axed in 1995 for political reasons. This is a rare example of the need for a prescriptive standard (that specified the additive) instead of a "performance standard." In the meantime, badly needed energy legislation has been delayed by the politics of the situation (oil companies want to be shielded from MTBE based liability).

The flammability of mattresses has long been a major safety concern. California legislated Bill 603 in 2001 that was effective in 2002. The bill made use of two ASTM standards for flammability testing D7016 and D7140. In another example of state leadership, the resulting state standard became the basis for a federal standard of the Consumer Product Safety Commission (CPSC) 16 CFR Part 1633 (Wilhelm 2007).

States' Departments of Agriculture

Most states have a Department of Agriculture that looks after agricultural issues, including standards and conformity assessment. The State Departments of Agriculture complement the U.S. Department of Agriculture's work to improve the health and safety of animal and plant products. One of the functions of some such departments is legal metrology (including weights and measures). State metrology laboratories use "measurement standards" that are traceable to NIST and eventually to the International Organization for Legal Metrology (OIML). Thus, the liter of gasoline that you buy is traceable to an International Standard! The weights and measures activities include the monitoring of gasoline dispensers to assure that liquid measuring at the pump is accurate. Gasoline pumps frequently bear seals certifying their accuracy. Moreover, the *quality* (octane rating) of the gasoline dispensed is part of the conformity assessment process at the state level (Opperman 2005).

One of the more prominent State Departments of Agriculture is the Pennsylvania Department of Agriculture (PDA). Pursuant to the Pennsylvania Bakery Law, the "Reg. P.A. Dept. of Agr." appears on many food products and certifies that the product is clean and wholesome. Pennsylvania licenses its Certification mark to out-of-state bakeries, which are monitored through contract with other state agriculture departments. The Commonwealth of Pennsylvania is an exporter and importer of food and food services of various kinds. It promotes the stewardship of the land and the forests. The PDA works to prevent environmental damage from nutrients (fertilizers) used in agriculture. The department is located in the commonwealth capital of Harrisburg.

States Have Options on ICT Standards

States as well as the federal government have great influence as customers for newly emerging information and communication technology (ICT) standards. Because many existing ICT systems "grew like Topsy" in an uncoordinated manner, the coordination of standards at the state level could result in significant savings (Johnson 1995). Johnson recommends an architecture framework for the National Information Infrastructure (NII) and a Global Information Infrastructure (GII) to enhance communication nationally and internationally. Johnson also reports how California legislated to make government information available online. A federal Government Information Locator Service (GILS) has been proposed. The patchwork of conflicting regulations is also cited as an impediment to the adoption of modern standards for communications.

Four states have, or are considering, the mandating of Open ICT Standards for government agencies: California, Massachusetts, Minnesota, and Texas (Updegrove 2007). The legislation proposed in Texas would require that an Open Document Format (ODF) be used for state electronic documents that meet the following requirements:

1. Interoperable among diverse internal and external platforms and applications
2. Published without restrictions or royalties
3. Fully and independently implemented by multiple software providers on multiple platforms without any intellectual property reservations for necessary technology
4. Controlled by an open industry organization with a well-defined inclusive process for the evolution of the standard

References

Fewell, Roy, 1997, International Conference of Building Officials (ICBO), *ASTM Standardization News* 25(1): pp. 20–25.

Herman, Marilyn, and Wendy Clark, 2002, ASTM's New Research Report on Reformulated Gasoline, *ASTM Standardization News* 30(6): 26–29.

Johnson, Jerry L., 1995, Standards for the Information Infrastructure: Barriers and Obstacles, in *Standards Policy for Information Infrastructure*, edited by Brian Kahin and Janet Abbate, Cambridge, Massachusetts, MIT Press, pp. 502–515.

Johnson, Mark A., 2003, Referenced Standards and the International Codes, *ASTM Standardization News* 31(12): 34–37.

Kandhal, Prithvi S., 2003, History of ASTM Committee D04 on Road and Paving Materials (1903–2003), *ASTM Standardization News* 31(9): 20–25.

McNichol, Dan, 2003, *The Roads That Built America*, New York, Sterling.

National Institute of Standards and Technology (NIST), 2007, NIST Recommendations in New Model Building Code, *SES Standards Engineering* 59(4): 9.

Opperman, Henry, 2005, NIST's Role in Weights and Measures, *ASTM Standardization News* 33(1): 34–37.

Scheibal, Stephen, and Ralph K. M. Haurwitz, 2001, East Austin's Soiled Earth, *Austin American-Statesman*, March 3, pp. 1,10.

Texas Commission on Fire Protection, 1992. Title 37. Public Safety and Corrections, Part XIII, *Texas Register*, vol. 17, p. 8504 et seq.

Thompson, Robert C., 2002, New York State: Building a Case for Standards, *SES Standards Engineering* 54(6): 12–16.

Updegrove, Andrew, 2007, Meanwhile, Deep Down in Texas: An Open Format Bill is Filed. *Consortium Standards Bulletin* February 1, pp. 17–22.

Wilhelm, Richard, 2007, Protection While We Sleep, *ASTM Standardization News* 35(9): 56.

10

Legal Issues in Standards and Conformity Assessment

Introduction

A variety of legal issues surround the subjects of standards and conformity assessment. The object of this chapter is not to provide legal advice but merely provide some engineering perspectives on some of those subjects. Readers who want legal guidance on any of these issues are advised to consult an attorney. It is hoped the perspectives developed here will enhance the communication of the reader with an attorney. Some of the legal cases described here, for example, *ASME v. Hydrolevel* and *Allied Tube and Conduit Corporation v. Indian Head*, are only mentioned in passing in many books on standards. These and other issues have exerted such a profound influence on modern standards development and use that they warrant coverage in this book. Products liability involving products covered directly by special Federal Regulations on safety, such as those for consumer products, are not covered here. Readers desiring standards information on consumer products are referred to 16 CFR 1000, the Regulations of the Consumer Product Safety Commission.

Products Liability of Manufacturers and Others

The question frequently arises "Does adherence to a standard help avoid liability for a manufactured product?" If one asks an attorney, the answer might be it depends. Although there are some interesting connections between "mandatory standards" (Regulations) and products liability, the focus here is on Standards, including those that are "incorporated by reference" in federal, state, and local codes such as the National Electrical Code. Products liability litigation takes place mostly at the state level so some of the variability of the subject is due to variations in products liability laws among the states. Some of the variability from state to state has been reduced by the Model Uniform Products Liability Act (MUPLA) in states that have adopted it at least in part (U.S. Department of Commerce 1979). The flavor of MUPLA, which is based on products liability case law, is evident in the following excerpts:

> The Model Law, if enacted by the states, would introduce uniformity and stability into the law of product liability. This, in turn, would help stabilize product liability insurance rates. Uniformity and stability in this area are needed because product liability insurance rates are set on a countrywide basis. Thus, product liability law differs from medical malpractice, automobile and other standard lines of liability.

The Model Uniformity Products Liability Act addresses the issues regarding the value of adherence to a standard in defending a products liability action in Section 107: Relevance of Industry Custom, Safety or Performance Standards (pp. 62729 et seq.). A relevant paragraph reads in pertinent part as follows:

> Compliance with standards that are rigorous and objective [in that they were developed through careful, thorough product testing and a formal product safety evaluation, and up-to-date in light of the technological and scientific knowledge reasonably available at the time the product was manufactured] suggests that a product was *not defective*. Failure to comply with such standards, especially performance standards, suggests that it was defective. (emphasis added)

Section 108: Relevance of Legislative or Administrative Regulatory Standards and Mandatory Government Contract Specifications follows a similar line of reasoning. It is generally established that the failure of a component, for example, a fastener, that results in injury produced by an equipment using that component, makes both the end equipment manufacturer *and* the fastener maker liable. Also, every party in the chain of commerce is potentially liable for a defective product including sellers and distributors. A good, but dated, introduction to products liability appears in *What Every Engineer Should Know About Product Liability* (Thorpe and Middendorf 1979). Much of the basic terminology of the law on products liability is explained therein. Pursuant to the Uniform Commercial Code (UCC), legal theories of products liability include express warranty, implied warranty, *negligence*, fraud and, *strict tort liability* (Twomey et al. 2002). The impact of technical standards on products liability is mostly applicable to cases of negligence and strict liability. The interaction of safety standards and expert witnesses is very important in terms of admissibility of evidence, the determination of the state of the art, and other aspects of products liability litigation. Ruggieri (2004) provides an accessible treatment of the admissibility of expert witness testimony and applicable standards.

Traver (1982) reported how a failure to follow a standard has been used to show negligence in a products liability case. The defendant's power company violated the National Electrical Code by installing Edison base type receptacles. The use of standards in *defending* products liability cases is more problematic. It is treated by Hansen (1996) who states that industry standards are more applicable to "design defect cases" than to other theories of liability. Whether based on negligence or strict liability, most cases on product liability involve design defects, manufacturing defects, or failure to warn. The several theories of product liability are not mutually exclusive; the plaintiff may, depending on the facts of the case, have several theories of action. The potential applicability of various standards to the issues follows.

Design Defects

The use of a widely recognized product safety standard, especially one that is American National Standards Institute (ANSI) recognized and included by reference in state or local codes, can be used in court to rebut the allegation that a design *defect* exists in the accused product. Courts frequently hold that the product safety expertise embodied in such a standard carries more weight than the opinion of a coterie of self-appointed experts. A plaintiff's attorney will likely focus on the goodness of such a standard (see Chapter 3). The standard must be up to date and preferably harmonized with International Standards on the same subject where they exist. Product designers should not view a product safety standard as a shield from liability; they should consider it a tool to be used to develop safe and marketable products. Most standards are minimal requirements. Designers should carefully consider if the product's safety can be enhanced by *exceeding* the standard's requirements, especially with regard for predictable misuse (unpredictable misuse if sometimes a valid defense).

Manufacturing Defects

Manufacturing defects can be minimized by the intensive use of a good quality-system-based manufacturing standard, such as an ISO 9000 family standard, especially where the manufacturer's quality system is *accredited*. Reliance on a good *process* is the key to minimizing manufacturing defects. Detailed work instructions and a culture that promotes defect-free manufacturing are essential. Detailed scrutiny of all of the components that go into a product is essential (conformity assessment). Hunter (1992) has recommended the use of only UL Recognized Components in electrical equipment as a *configuration management* tool. In the case of the UL Certified product manufacturer, the Follow-up Service Procedure becomes a vital part of the manufacturing quality standard. Frequent internal audits, combined with external audits, minimize the variations in products that affect safety (see Chapter 2).

Failure to Warn

Failure to warn is a major part of much products liability litigation. Suitable warnings for foreseeable hazards are, of course, most important. Product safety standards frequently stipulate warnings in certain situations but consideration should be given to "special warnings" regarding less commonly perceived hazards. A general principle is that a warning should never be used to avoid a reasonable design change that could eliminate the hazard. Instructions and warnings are usually required to be given in the language of the user (for example, in both French and English in Canada and in both English and Spanish in the United States). How to effectively warn in multiple languages is an issue that gets some help from the IEC 60417 standard (Graphical symbols for use on equipment).

McDonnell and Miller, Inc. Caper, a.k.a *ASME v. Hydrolevel*; Federal Trade Commission (FTC) Notice of Proposed Rulemaking (NPRM) on Standards and Certification

These issues substantially overlapped, both in time and in subject matter, and so are being treated together here. The *ASME v. Hydrolevel* case was not about standards setting per se, as is sometimes reported. It was, in fact, about the *interpretation* (a loose form of conformity assessment) of an American Society of Mechanical Engineers (ASME) Boiler and Pressure Vessel Code (see Chapter 8). Tekla S. Perry (1982) provides a good chronological summary of the story. McDonnell and Miller, Inc., (M&M) of Chicago had long dominated the market for fuel cutoffs used to prevent explosions in boilers when the water was too low; that traditional technique was an arrangement that cut off the boiler fuel when the water level fell below the lowest visible part of the glass water gauge. An upstart competitor, Hydrolevel Corporation of Farmingdale, New York, placed on the market a competitive fuel cutoff device that differed structurally from the M&M fuel cutoff in that it used an electric probe and incorporated a time delay before the fuel was cut off because the variability of the water level could otherwise result in nuisance cutoffs. Hydrolevel won an important major customer from M&M and threatened take market share away from M&M.

Several officers of M&M, one of whom was a vice chair of the relevant ASME Boiler and Pressure Vessel subcommittee, hit upon a scheme to eliminate Hydrolevel's competition. They wrote a letter to the Boiler and Pressure Vessel subcommittee asking for an interpretation on whether the Hydrolevel cutoff complied with the relevant code requirements. One of the officers then unethically used his position on the Boiler and Pressure Vessel subcommittee to state, without full subcommittee approval, that Hydrolevel's cutoff (probe) device did not comply with the code requirements. That interpretation was later modified by the full ASME subcommittee but the result was a powerful disincentive for potential customers to purchase the Hydrolevel product. Hydrolevel's ensuing problems were flagged by a *Wall Street Journal* article, which resulted in a vetting of the issue by the U.S. Senate Subcommittee on Antitrust and Monopoly in 1975. Hydrolevel sued ASME, M&M's new owner International Telephone & Telegraph (ITT), and the Hartford Steam Boiler Inspection and Insurance Co., which had participated in the caper, for violating the Sherman Antitrust Act (15 U.S.C.). ITT and Hartford settled before the trial in the U.S. District Court, Eastern District of New York, so the suit proceeded against ASME only. The jury found in Hydrolevel's favor; ASME was liable under the Sherman Act and subsequently assessed treble damages of $9.9 million. ASME appealed the decision to the Second Circuit Court of Appeals.

FTC Notice of Proposed Rulemaking on Standards and Certification (43 FR 57269 of December 7, 1978)

While *Hydrolevel* was winding its way through the federal court system, the publication of the NPRM and accompanying staff report (FTC 1979) hit the standards and certification community like a bomb. Armed with the lower court's *Hydrolevel* decision and a clutch of other examples of alleged restraint of trade violations by standards developers and certifiers, the FTC proposed to overlay the existing standards and certification system with an infrastructure of legalistic regulatory requirements that would have essentially federalized the system. Although the legal–economic issues were well presented, the staff report reflected a substantial naiveté on the practical aspects of product safety and product liability. For example, the FTC attacked the results of ASTM small-scale standard test methods of foam plastics used as insulation in buildings. The FTC apparently didn't appreciate the concept that you cannot, in many cases, conduct a realistic test for practical reasons (e.g., one can't burn down a large building to provide a more realistic test). Also, the FTC recommendation that graded *safety* requirements that permit buyers to buy products that are less safe than others would have subjected manufacturers and others to many products liability lawsuits. There were many other impractical concepts proposed in the NPRM, some of which were critiqued in publications (e.g., Hunter 1980 and Sherr 1982). The staff report bemoaned the fact that small producers could not afford (then) to participate in standards setting but ignored that they reaped an economic benefit as "free riders" in the process, which (theoretically) should have made them more competitive in the martketplace. The staff report also attacked consensus and balance concepts because they do not always produce the correct result.

The NPRM galvanized the standards and certification community into action against it. Most of the large bodies presented arguments against the proposed rule in subsequent hearings. ANSI and fifteen other organizations unsuccessfully sued the FTC in 1979 alleging the rulemaking proceedings violated certain constitutional and statutory rights. Nonetheless, standards community organizations began to scrutinize and improve their procedures to prevent the anticompetitive activities identified by the FTC, which became a principal impetus for the movement to "due process" in standards development.

Further efforts to issue a rule were effectively terminated by the Federal Trade Commission Improvement Act of 1980 (P.L. 96–252) under the Reagan administration.

The ASME appealed the district court's decision to the federal appeals court, which decided that ASME was liable for antitrust law violation based on the "apparent authority" of its members (agency law in legal speak) and the intent of the Sherman Act. Pursuant to that action and a cross appeal by Hydrolevel, the appeals court did find the damages award excessive. ASME then took the case to the Supreme Court (hence the name *ASME v. Hydrolevel*).

The Supreme Court essentially agreed with the decision of the appeals court as follows:

> Petitioner, the American Society of Mechanical Engineers, Inc. (ASME), is a nonprofit membership corporation organized in 1880 under the laws of the State of New York. This case presents the important issue of the Society's civil liability under the antitrust laws for acts of its agents performed with apparent authority. Because the judgment of the Court of Appeals upholding civil liability is consistent with the central purposes of the antitrust laws, we affirm that judgment.

The Supreme Court went on to say:

> Furthermore, a standard-setting organization like ASME can be rife with opportunities for anticompetitive activities. Many of the ASME's officials are associated with members of the industries regulated by ASME Codes. Although, undoubtedly, most serve ASME without concern for the interests of their corporate employers, some may well view their positions with ASME, at least in part, as an opportunity to benefit their employers. When the great influence of ASME's reputation is placed at their disposal, the less altruistic of ASME's agents have an opportunity to harm their employer's competitors through manipulation of ASME's codes.

As a result of the Supreme Court's decision in *ASME v. Hydrolevel*, which was published by IEEE in September 1982, standards developers intensified their efforts to improve their procedures for standards (and interpretations) to minimize the probability that a similar action would be repeated. The development of interpretations became subject to the same rules for balance and fairness as those for standards development (see Chapter 4). ANSI reorganized its structure, spinning off the standards developing components of the organization, which were separately incorporated. Some of the basic problems for standards setters posed by *Hydrolevel* might have been avoided or minimized if the ASME code had been more performance oriented rather than descriptive.

The FTC issued a final staff report in April 1983 summarizing the proceedings, which included nearly 1,000 written comments, and pointing out that the U.S. Supreme Court's decision in *Hydrolevel* should further discourage standards developers and certifiers from abusing the antitrust laws (FTC 1983). The final staff report recognized that steps had been taken by most standards organizations to improve their procedures with the exception of the handling of complaints and stated that *Hydrolevel* and, perhaps, other events since the close of the record on January 1980 may affect the need for a rule. The result was that the FTC returned to case-by-case actions to prevent abuses of the FTC Act.

Indian Head, Inc. v. Allied Tube and Conduit Corporation and the National Fire Protection Association (NFPA), a.k.a. Allied Tube and Conduit Corporation v. Indian Head, Inc.

Carlon, a division of Indian Head, Inc., manufactured a plastic (PVC) conduit to carry electrical wires but could not, as a practical matter, market it successfully absent an approval in the National Electrical Code (NEC) of the National Fire Protection Association (NFPA), which had approved a steel-based conduit. Carlon sought to get the PVC conduit approved in the 1981 edition of the NEC on the basis that it was more pliable, cheaper, and less likely to short circuit conductors than the steel conduit in use. The action arose from Allied's efforts to prevent the adoption of PVC conduit in the NEC based on the fact that, in a building fire, burning PVC conduit would emit toxic smoke and gas (chlorine) and the apparent fact that approving PVC conduit would damage Allied's dominance of the conduit business.

The NFPA approves periodic changes (every three years for the NEC) to its codes by having changes developed by Code Making Panels (CMPs) of experts operating in the consensus standardization mode and vetting the changes publicly. Public comments are considered by the CMP and incorporated in the code provisions as considered appropriate. The changes are then submitted to a majority vote at an annual meeting of all NFPA members for final approval. As mentioned earlier, codes like the NEC are incorporated into state and local electrical codes unchanged or with modifications.

To prevent the adoption of the PVC conduit in the NEC, Allied and other elements with steel interests agreed to bring as many NFPA voting members to the annual meeting as possible. Allied paid more than $100,000 to various nonexpert individuals to join the NFPA, including employees, sales agents, the agents' employees, company executives, employees of two Allied divisions, and the wife of the national sales director.

Allied also covered expenses for the newly minted NFPA members to attend the annual meeting in Boston (in the spring of 1991). Other steel interests also sent new NFPA members to the annual meeting, resulting in a total of 230 new votes in opposition to the adoption of PVC conduit. At the annual meeting, Allied instructed the new members where to sit, and how and when to vote. Few of Allied's voters had any of the technical documentation necessary to follow the meeting. The adoption of PVC conduit was narrowly defeated by a 394 to 390 vote. Carlon appealed the decision to the NFPA Standards Council, which passed it on to the Board of Directors. The Board of Directors found that the NFPA rules had been circumvented but not violated, thus returning the proposal to the CMP. The NFPA has a method, called a Tentative Interim Amendment (TIA), for minor changes to the code during the regular three-year revision cycle to accommodate "emergency" changes. Carlon (Indian Head) then requested a TIA but that, after undergoing the usual process, was denied at least in part by Allied's continuing opposition.

In October 1981, Carlon sued both Allied and the NFPA, alleging that Allied and its agents had "combined and conspired to prevent competition from nonmetallic tubing by excluding it from the NEC *by subverting the consensual standards rules and procedures of the NFPA*" (emphasis added). Indian Head discontinued the suit against the NFPA before the trial began in the U.S. district court. The jury found for Indian Head but the judge set aside the $3.8 million verdict on the basis that Allied was "lobbying legislatures" indirectly via the NFPA, a protected activity of First Amendment cases called the *Noerr-Pennington* doctrine. Indian Head appealed that decision to the United States Court of Appeals for the Second District. Attorneys from the Federal Trade Commission and the Department of Justice appeared at the appeals court as amicus curaie (friends of the court). The appeals court reversed the district court's holding and reinstated the $3.8 million, $11.4 million after trebling, against Allied, citing *Hydrolevel* along the way.

Allied appealed the appeals court verdict to the U.S. Supreme Court in 1987. The Supreme Court upheld the appeals court decision (7–2 vote). The Court noted that:

> private certification laboratories, such as Underwriters Laboratories, normally will not list or label an electrical product that does not meet code standards; many underwriters will refuse to insure structures that are not built in conformity with the code; and many electrical inspectors, contractors and distributors will not use a product that falls outside the code.

The minority opinion worried about the effect on standards bodies and noted:

> There is no doubt that the work of these private organizations contributes enormously to the public interest and that participation in their work by those who have the technical competence to do so should not be discouraged. The Court's decision today will surely do just that.

The standards community was shocked at the size of the Indian Head award and wondered if any reasonable procedural rules of a standards developer could avoid the abuse of the system by a determined party. The importance of *rationale* for standards provisions was reemphasized and SDO procedures were again tightened. Howard I. Forman (1997), an attorney and recipient of the 1981 Leo B. Moore Medal of the Standards Engineering Society, referenced the *Hydrolevel* and *Allied Tube* cases in his paper that advocated the potential use of Alternative Dispute Resolution (ADR) in reducing the litigation in such cases. Attorney David A. Swankin (1990) made studied uses of the *Hydrolevel* and *Allied Tube* cases as well as the FTC Proposed Rule in outlining due process procedures (see Chapter 4) for reducing the risk of antitrust liability. Cliff Edwards (2007), outlining recent developments in WiMax standard (802.20) by the IEEE, reported alleged, *but not proven*, attempts by Qualcomm to pack the committee in a manner similar to the *Allied Tube* case (Edwards 2007). The committee deadlock, produced by Qualcomm's opposition to the new standard, was resolved when IEEE changed the procedure to require that "all votes and ballots in the working group be cast on the basis of entity affiliation, with a single vote for each entity" (*IEEE-SA News*, July 31, 2007). Another interesting case involved ASTMI (Brooke 1995). Brooke points out that the FTC, as part of its review in the NPRM on Standards and Certification, held up the ASTMI standard development process as a model for others to follow.

National Cooperative Research Act (NCRA) of 1984 (P.L. 98-462)

As mentioned in Chapter 6 in the discussion of standards setting consortia, the NCRA was the "enabling act" for standards setting consortia. The NCRA provides that a person's conduct in making or performing a contract to carry out a joint research and development venture shall not be deemed illegal per se under Federal antitrust laws (or similar state laws) but shall be judged on its *reasonableness* (emphasis added) considering factors including its effect on competition in the relevant markets. The "joint research and development venture" excludes parties that resort to anticompetitive behavior. The joint research and development venture liability under antitrust laws is limited to actual damages and the cost of the suit. The subject joint ventures, in order to realize the treble damages immunity, are required to file notifications to the Attorney General and the Federal Trade

Commission. The filings must identify the parties involved and the nature and objectives of the venture. There also some other requirements for the immunity (no bad faith, etc.). The notifications are published in the *Federal Register* so that the operation of the joint venture becomes rather transparent to the public as well as antitrust authorities. Andrew Updegrove (1995) has pointed out some disadvantages of that transparency. The National Cooperative Research Production Act (NCRPA 1994) [P.L. 1032-42] extended the same detrebling of antitrust damages to joint ventures performing *production* as well as research and development. As a result of the NCRA, there was an explosion of standards setting consortia beginning in the mid-1980s, mostly in the software area.

Stephen Oksala (2000) credits the NCRA, the EU's New Approach (Chapter 7) and the WTO TBT Agreement with profoundly changing the standards world.

Standards Development Organization Advancement Act (SDOAA) of 2004 (P.L. 108-237)

The U.S. president signed the SDOAA (previously known as the Standards Development Organization Advancement Act of 2003) into law in 2004. This act extended the provisions of the NCRPA to protect Standards Development Organizations from the treble damages (leaving them liable for ordinary damages) of the antitrust laws if they file notices with the Department of Justice and the Federal Trade Commission advising them of the scope of their activities. SDOs are thus subject to a "rule of reason standard" regarding their liabilities under the antitrust laws. This legislation advances the ability of SDOs to work with the federal government in the development of standards suitable for government use, as envisioned in the National Technology Transfer and Advancement Act (NTTAA) of 1995 (see Chapter 11 for a discussion of the NTTAA). The legislation extends to individual SDOs the benefits of relief from treble damages in the same manner as the National Cooperative Research Act (NCRA) of 1984 described earlier.

Civil Liability of Standards Developers

There have been several instances of civil litigation against SDOs, including one called *Meneely v. S. R. Smith, Inc. et al.* The Washington State Supreme

Court found a standards developing trade association, the National Spa and Pool Institute (NSPI), to be liable for injuries to a sixteen-year-old boy who dove into a swimming pool. The court found that the NSPI failed to adjust its standard in view of a recognized hazard and did not *warn* consumers of the recognized hazard. The court's decision was based on a legal theory called the "voluntary rescue doctrine." Some legal experts questioned the applicability of the voluntary rescue doctrine to the facts of the case. The $6.6 million judgment bankrupted the NSPI (Purcell and Fox 2001). Purcell and Fox emphasize the lesson that developers of safety standards should take care to insure that the standards they develop are up to date and take into account all known hazards and promptly warn the public of newly discovered hazards (duty to warn). If the defective standard had been an International Standard, a broader range of expertise might have been used to avoid the hazard.

A "legislative remedy" for standards developers in such cases appears in the *Standards Developing Organizations Advancement Act* of 2006. The act protects SDOs that meet ANSI due process requirements from civil liability except for "intentional or bad faith misconduct." The 2006 SDOAA preempts state and federal actions for liability except for the exceptions indicated.

It is interesting to note that the Federal Trade Commission has obtained a measure of control over the standards community via the legislation described in the previous sections. The legislation is published in the U.S. Code, 15 USC Sections 4301 through 4306. The legislation described permits the full realization of the National Technology Transfer and Advancement Act (NTTAA) described in the next chapter.

Standards and Intellectual Property Rights (IPRs)

Intellectual property rights include patents, copyrights, trademarks, trade secrets, and other issues. Only patents and copyrights have a substantial intersection with standards (except for the trademark issues that prevented Java™ from becoming an International Standard), so they will be treated here. Those wanting a good introductory treatment of patents are referred to *Nolo's Patents for Beginners* (Pressman and Stim 2006). Those desiring a good introductory treatment of copyright are referred to *The Copyright Handbook* (Fishman 2004). Because both patent rights and copyright are constitutionally protected monopolies, there is a substantial tension, to say the least, between these subjects and standards development. These subjects also affect the economics of standards and conformity assessment, a subject that is treated in Chapter 12. Oliver Smoot (1995) provides a good survey of the interaction of standards and intellectual property rights, including the situation with regard to the International Standards bodies ISO and IEC as well as the European Standards body ETSI.

Patents and Standards

Patents and standards are related but opposing concepts; both are vehicles for technology transfer. Historically, little thought was given to patents when developing standards. One exception was the IEEE adoption of the famous Digital Interface for Programmable Instrumentation (IEEE Standard 488). The Hewlett-Packard Co., a major maker of instrumentation, brought its Company standard to the IEEE for adoption. IEEE decided to adopt the standard and endorse it to ANSI on the following condition. "The Hewlett-Packard Company has assured the IEEE that it is willing to grant a license under these patents on reasonable and nondiscriminatory terms and conditions to anyone wishing to obtain such a license" (IEEE 1975). As this writer recalls, the license cost $100. IEEE Std 488-1975 was a very important standard that was coordinated with IEC standardization efforts. IEEE Standard 488 was described in the literature by Hewlett-Packard authors that pointed out the synergism of the 488 interface and microprocessors (Loughry and Allen 1978). The development of IEEE 488 gave birth to the "virtual instrument" industry of which Austin-based National Instruments is a leading example (the best $100 investment they ever made). The IEEE 488 standard was harmonized with IEC 625-1 and IEC 625-2 in 1978 corresponding to two complementary standards, IEEE 488.1 and IEEE 488.2. Another historical case of the intersection of patents and standards was the Federal Communications Commission (FCC) requirement that Western Electric, the licensing arm of AT&T, grant a royalty bearing license to all future manufacturers and users of the famous modular plug and jack system (see Chapter 3 and 47 CFR 68.504 [10-1-93 Ed.]).

The Patent Problem for Standards Developers

No standards body wants to publish a standard that cannot then be implemented without paying royalties for a license under a patent if it can be avoided. If a standard involves a state-of-the-art technology, it is likely that a patented (new and useful) invention will be involved. If the patent covers a basic invention, it may be nearly impossible to find an alternative for standards development that is not technically inferior to the patented solution. If the patent is an improvement patent, the standards developers can frequently find an alternative solution to the problem *if they find out about it before publishing the standard*. So the timing of the SDO's discovery that a patent is involved is critical. Steven J. Frank (2002) points out that the patent licensing terms for such situations should be as transparent as the SDO's development process. The European Telecommunication Standards Institute (ETSI) early on tried to enforce a form of compulsory licensing on its standards participants (Shurmer and Lea 1995). The result was a wall of opposition from large holders of patents, especially in the United States. ETSI finally modified its policies to something that was acceptable. Some of the consortia including the W3C and others like the IETF similarly tried to severely limit the inclusion of patented technology in their standards but that was also found to be untenable. Some large patent holders continued to game the system to maximize patent royalties instead of cooperating to develop standards. The completion of the GSM cell phone standard was reportedly delayed for three years while patent issues were being thrashed out (Hjelm 2000). Several authors have argued for a "rebalancing" of IPR issues with Standards that improve social welfare, for example, interoperability standards (Farrell 1995; Egyedi 2001). It was announced in April 2007 that the ISO, IEC, and ITU had harmonized their patent policies.

Patent Ambush

Patent ambushes come in two different scenarios: the *active patent ambush* and the *passive patent ambush*. In the active patent ambush, the patent holder participates in the SDO process but does not disclose its applicable patent holding in a timely manner, that is, before the standard is developed, issued, and widely implemented. Patent holders with large portfolios sometimes claim ignorance of the relevant patent until it is too late for the SDO to take evasive action. If the patent holder's action is intentional, it is not only unethical but possibly illegal (under the Sherman or FTC Acts). In some cases, accused patent infringers have used that argument as a defense. Brief descriptions of alleged active patent ambushes are the cases of Dell, Unocal, and Rambus described in the following.

FTC v. Dell Computer

A Dell engineer, serving on the VESA Local Bus Standards working group, signed a statement that Dell held no IPR in that effort. It was the position of the VESA working group that no proprietary technology would be used since some of the technical alternatives were in the public domain. Later, after the standard was set, Dell began to assert its patent covering the VESA Local Bus and was then sued by the FTC, alleging that Dell intentionally hid its patent on the VESA technology in order to profit from subsequent licensing fees. In any event, Dell lost the patent in question.

FTC v. Unocal

Air pollution regulators in California met in the 1990s with a group of oil companies to find a way to reduce the pollution of auto emissions (reformulated gasoline; see Chapter 9). While the parties expected that the solution would be in the public domain, Unocal quietly obtained patents on the state standardized means of reducing air pollution and allegedly tailored the patent claims to read squarely on the resulting solution (Barrionuevo

2000). After the regulations had been set, Unocal sued the other oil companies under its patents, resulting in a protracted litigation. In 2003, the FTC sued Unocal for anticompetitive practices (Banerjee 2003). This case differs from Dell in that the standard setter was the California Air Resources Board rather than a private standards group. Because of the FTC's action and other factors, Unocal forfeited its portfolio of patents on reformulated gasoline to settle the FTC complaint.

In the passive patent ambush, a patent holder does not participate in the SDO process but waits until the standard is published and the industry commits to the implementation of the standard. If the patent had not been issued before the standard was finalized, the patent is sometimes called a *submarine patent*. A real submarine patent is often held to be one that languished in the Patent and Trademark Office (PTO) files under an extended prosecution that may involve divisional and or continuing applications that retain the priority date of the original application (Heinze 2002). Opportunities for that kind of operation are being reduced by changes and contemplated changes in the patent laws. Faux submarine patents are those not "discovered" until the standard has been issued and implemented. Because of a quirk in the patent law, "ignorance is bliss" on the part of an infringer; if an infringer admits knowledge of a patent that it later infringes, it could possibly be liable for treble damages.

Technical Complexity and ICT Convergence Complicate IPR/Standard Issues

The complicated issues described here have been dubbed the *patent thicket*. The problem of cross-licensing before the standard is developed is severe under normal circumstances. There is also the formidable problem of "hold-up," an infringement of a patent issued after the standard is set. A variety of possible solutions to such problems is the formation of a patent pool where members of the standards body, frequently a consortium, can work out the details of equitable cross-licensing *before* the standard is developed. Of course, the FTC and Department of Justice take an active interest in such arrangements, especially if a complaint is filed, and even approve such patent pools providing certain criteria are observed.

Patent pools are not new; one of the earliest being the aircraft patent pool (1917), resulting from a recommendation of Franklin D. Roosevelt, then assistant secretary of the Navy. More recent examples include the IEEE 1394 (Firewire) Standard (Clark et al. 2000). The 1394 patent pool includes thirty-four patents that are licensed to fifty-six licensees on a worldwide nonexclusive and nontransferable basis at a cost of $0.25 per use.

The Rambus Story (Active Patent Ambush)

The first part of the Rambus story is detailed in a readily accessible reference (Jaffe and Lerner 2004). Rambus, a small company that designed computer memory systems, filed a patent application in 1990. A number of divisions and continuations took place over nearly a decade resulting in several patents on variations of the Random Access Memory Bus (hence the name Rambus). Rambus's basic invention permitted computer units to access Dynamic Random Access Memories (DRAMs) more quickly than previous circuitry (Kumagai 2001). Meanwhile, Rambus representatives sat on the Joint Electronic Device Engineering Council (JEDEC; see Chapter 8) standards committee that was formulating standards of DRAMs and successive improvements such as SDRAMs and DDR SDRAMs. JEDEC required that committee members disclose pending patent applications and agree to license their patents at reasonable royalty rates. JDEC committee members were unaware that Rambus had filed the divisional and continuation applications, obtaining the priority date of the original application on which a patent had been issued and of which the committee was aware.

After several of its additional patents were issued and the semiconductor industry had commenced production of the improved SDRAM chips, Rambus sued Infineon, a major producer of those chips, for infringement of its patents in 2000. Infineon countersued in federal court in Virginia, charging Rambus with fraud because of Rambus's activity on the JEDEC committee. The issue was appealed to the federal appeals court that sided with Rambus (Parloff 2003). The issue revolved around the *sufficiency of the JEDEC committee rules*. All of this activity, of course came to the attention of the FTC, which in 2003 charged Rambus under a different law (deception under Section 5 of the FTC Act). An August 2, 2006, FTC press release stated "FTC Finds Rambus Unlawfully Obtained Monopoly Power." The press release continues:

> The Commission's unanimous opinion states, "We find the Rambus's course of conduct constituted deception under Section 5 of the FTC Act. Rambus's conduct was calculated to mislead JEDEC members by fostering the belief that Rambus neither had, nor was seeking, relevant patents that would be enforced against JDEC-compliant products ... Under the circumstances, JEDEC members acted reasonably when they relied on Rambus's actions and omissions and adopted the SDRAM and DDR SDRAM standards."

A subsequent FTC press release of February 5, 2007, reads "FTC Issues Final Opinion and Order in Rambus Matter." The result was a cap on the royalties that Rambus could charge for SDRAMs and DDR-SDRAMs. The result of the FTC action on the Rambus matter was yet another wake-up call to standards developers to make sure that their "patent provisions" were adequate and less capable of being "gamed." The Rambus story is apparently not over. In April 2008 there was a *Wall Street Journal* report that a federal appeals court overturned the FTC ruling against Rambus (Clark and Richmond 2008).

Analysis of Reasonable and Nondiscriminatory Rates (RAND) in Licensing

Like the Hewlett-Packard license on its IEEE 488 bus patents described earlier, SDOs usually require that each participant agree to license every applicant on a *reasonable and nondiscriminatory* (RAND) basis (Kipnis 2000). Questions that frequently arise include: What is reasonable? And exactly what is non-discriminatory? A royalty or other licensing contract consideration may or may not be reasonable, depending on the details of the situation, for example, the scope of the patent claims and how much of the total "machine" that it covers. For example, the holder of a patent on a special purpose "card" that goes into a PC may feel that the reasonable royalty should apply to the value of the whole PC, instead of just the value of the card. Historical royalties of 1 percent to 10 percent were not considered abusive in some situations. The patent holder may be a manufacturer who is more interested in cross-licensing than in collecting royalties. This considerably complicates the issue because it requires an evaluation of the patent portfolio that the manufacturer wants to cross-license. Feldman et al. (2000) provide an analysis of RAND provisions in IEEE and ITU. These and similar issues are likely best resolved in alternative dispute resolution proceedings as championed by Forman, rather than in courts of law where the process takes years to resolve and results in major delays for developing standards.

The efforts of some firms to game the system have required some SDOs like the IEEE to get more involved up front on the definitions of these arrangements: "The IEEE Standards Association (IEEE-SA) has enhanced its patent policy by permitting the disclosure of licensing terms associated with patents that might be included in its standards"(*IEEE-SA StandardsWire Online*, June 1, 2007).

Patents Clauses from The ANSI Essential Requirements: Due Process Requirements for American National Standards (January 31, 2006)

ANSI patent policy—Inclusion of Patents in American National Standards

There is no objection in principle to drafting a proposed American National Standard in terms that include the use of a patented item, *if it is considered that technical reasons justify this approach* [emphasis added].

If the Institute receives a notice that a proposed American National Standard may require the use of a patented invention, the procedures in this clause shall be followed.

STATEMENT FROM PATENT HOLDER

Prior to approval of such a proposed American National Standard, the Institute Shall receive from the identified party or patent holder (in a form approved by the Institute) either: assurance in the form of a general disclaimer to the effect that such party does not hold and does not currently intend holding any invention the use of which would be required for compliance with the proposed American National Standard or assurance that:

(a) a license will be made available without compensation to the applicants desiring to utilize the license for the purpose of implementing the standard; or

(b) a license will be made available to applicants under reasonable terms and conditions that are demonstrably free of any unfair discrimination.

RECORD OF STATEMENT

A record of the patent holder's statement shall be placed and retained in the files of the Institute.

NOTICE

When the Institute receives from a patent holder the assurance set forth in (a) or (b) above, the standard shall include a note as follows:

> NOTE—The user's attention is called to the possibility that compliance with this standard may require use of an invention covered by patent rights. By publication of this standard, no position is taken with respect to the validity of this claim or of any patent rights in connection therewith. The patent holder has, however, filed a statement of willingness to grant a license under these rights on reasonable and nondiscriminatory terms and conditions to applicants desiring to obtain such a license. Details may be obtained from the standard's developer.

RESPONSIBILITY FOR IDENTIFYING PATENTS

The Institute shall not be responsible for identifying all patents for which a license may be required by an American National Standard or conducting inquiries into the legal validity or scope of those patents that are brought to its attention.

Copyrights and Standards

With the exception of the Internet Engineering Task Force (IETF) and some other online standards, most standards are copyrighted by the standards bodies that produce them. Their copyrighted standards generate revenue for the SDOs, many of which obtain roughly half their funding from the sales of copyrighted standards. The price of some standards is relatively high since each party in the distribution chain gets a cut of the royalties. This inhibits more widespread use of these standards and has been a real impediment to education in standards (see Chapter 12). A reseller's point of view on digital rights management is expressed by Alison Ruger (2001).

Some legal experts have questioned whether standards should enjoy copyright protection (Samuelson 2006). One rationale is that standards are not an individual creation but are the product of a group of volunteers and hence should remain in the public domain. Court decisions have come down on both sides of the question, some holding that standards can be copyrighted and some that standards cannot be copyrighted. An argument has been made that software interface specifications should not be copyrighted, a view held in the EU, which provides for mandatory licensing of such standards (Warren-Boulton et al. 1995). Warren-Boulton et al. argue that de facto standards should not be afforded copyright protection.

One court decision that has recently concerned the standards community is known as *SBCCI v. Veeck* (ANSI 2003). SBCCI (Southern Building Code Congress International; see Chapter 9) is a nonprofit SDO that develops and promotes model building codes such as the Standard Plumbing Code, the Standard Gas Code, the Standard Fire Prevention Code, and the Standard Mechanical Code. These codes are "marketed" to state and local governments for use in their codes, subject to SBCCI's copyright, which controls sales of the basic standards. Veeck operated a nonprofit Web site that provided information about texts of local building codes. Veeck included the SBCCI code on his Web site without paying SBCCI a royalty.

SBCCI subsequently sued Veeck for copyright infringement in the U.S. District Court for the Eastern District of Texas, which found for SBCCI. Veeck then appealed that decision to the U.S. Court of Appeals for the Fifth District (*Veeck v. SBCCI* 1999). The appeals court reversed the district court opinion to rule that the SBCCI code did go into the public domain when made into law (by incorporation) by local legal action.

SBCCI appealed the appeals court decision to the U.S. Supreme Court, which, at least partially on the view of the U.S. solicitor general in an amicus brief, let the finding of the Fifth Circuit stand. ANSI's counsel thought that the U.S. Supreme Court might revisit the issue in a future case. If the *Veeck* case philosophy was extended to *all* copyrighted standards (including the National Electrical Code) "incorporated by reference" in federal, state and local regulations, it would destroy private standards setting *and* government regulation (see Chapter 11).

In summary, one might think that you would have to be an attorney to participate in a standards working group! That of course is inaccurate. Working group members should be aware of the legal boundaries of due process and consult with counsel, including patent counsel, if the situation warrants it.

References

American National Standards Institute (ANSI), 2003, *U.S. Supreme Court Decides Not to Adjudicate in SBCCI v. Veek: Standards copyright debate may continue on other fronts*, press release, July 8.

Banerjee, Neela, 2003, Unocal Is Sued by F.T.C. Over California Gas Patents, *The New York Times*, March 5, p. C6.

Barrionuevo, Alexei, 2000, Exhausting Feud: A Patent Fracas Pits Unocal Corp. Against Big U.S. Oil Producers; Smaller Firm Helped Shape New Air-Pollution Rules, Then Demanded Royalties, *The Wall Street Journal*, August 17, pp. 1, 8.

Brooke, Morris R., 1995, Federal Investigation of ASTM Standards, *ASTM Standardization News* 23(9): 26–29.

Clark, Don, and Riva Richmond, 2008, Court Overturns FTC Monopoly Ruling Against Rambus, *The Wall Street Journal*, April 23, p. B11.

Clark, Jeanne, Joe Piccolo, Brian Stanton, and Karin Tyson, December 2000, Patent Pools: A Solution to the Problem of Access in Biotechnology Patents? Washington DC, United States Patent and Trademark Office.

Edwards, Cliff, 2007, The Road to WiMax, *Business Week*, September 3, pp. 58–64.

Egyedi, Tineke M., 2001, IPR Paralysis in Standardization: Is Regulatory Symmetry Desirable? *IEEE Communications Magazine* 39(4): 108 –114.

Farrell, Joseph, 1995, Arguments for a Weaker Intellectual Property Protection in Network Industries, *ACM StandardView* 3(2): 46–49.

Federal Trade Commission (FTC), 1979, Standards and Certification, Proposed Rule and Staff Report, Washington DC, Federal Trade Commission, Bureau of Consumer Protection.

Federal Trade Commission (FTC), 1983, Standards and Certification, Final Staff Report, Washington DC, Federal Trade Commission, Bureau of Consumer Protection.

Feldman, Robert P., Maura L. Rees, and Brent Townshend, 2000, The Effect of Industry Standard Setting on Patent Licensing and Enforcement, *IEEE Communications Magazine* 38(7): 112–116.

Fishman, Stephen, 2004, *The Copyright Handbook*, 8th ed., Berkeley, CA, Nolo.

Forman, Howard I., 1997, A Sensible Alternative to Litigation, *SES Standards Engineering* 49(1): 4–6.

Frank, Steven J., 2002, Can You Patent an Industry Standard? *IEEE Spectrum* 39(3): 14–15.

Hansen, Steven W., 1996, Using Standards in Defending Product Liability Cases, *ASTM Standardization News* 24(2): 40–43.

Heinze, William F., 2002, Dead Patents Walking, *IEEE Spectrum* 39(5): 52–54.

Hjelm, Bjorn, 2000, Standards and Intellectual Property Rights: International Standardization of Third-Generation Mobile Systems, *SES Standards Engineering* 52(1): 10–16.

Hunter, Robert D., 1980, Future Directions in Federal Regulations, in *1980 IEEE Region V Conference Digest-Engineering and Society Interfaces, San Antonio, Texas*, pp. 49–51.

Hunter, Robert, 1992, Standards Enhance Competitiveness, *SES Standards Engineering* 44(6): 133–137.

IEEE, 1975, *IEEE Standard Digital Interface for Programmable Instrumentation*, adopted as American National Standard *ANSI MC1.1-1975*.

IEEE, 1982, Full Text of the U.S. Supreme Court Opinions in ASME v. Hydrolevel: Standards Abuse and Liability, *IEEE Technology and Society Magazine* 1(3): 3–12.

Jaffe, Adam B., and Josh Lerner, 2004, *Innovation and Its Discontents: How Our Broken Patent System is Endangering Innovation and Progress and What to Do About It*, Princeton, NJ, Princeton University Press, pp. 68–74.

Kipnis, Jason, 2000, Beating the System: Abuses of the Standards Adoption Process, *IEEE Communications Magazine* 38(7): 102–105.

Kumagai, Jean, 2001, Rambus: Friend or Foe? *IEEE Spectrum* 38(5): 42–47.

Loughry, Donald C., and Mark S. Allen, 1978, IEEE Standard 488 and Microprocessor Synergism, *Proceedings of the IEEE* 66(2): 162–172.

Oksala, Stephen, 2000, The Changing Standards World: Government Did It, Even Though They Didn't Mean To, *SES Standards Engineering* 52(6): 1–6.

Parloff, Roger, 2003, Technical Win for Rambus in Patents Case, *IEEE Spectrum* 40(4): 22–25.

Perry, Tekla S., 1982, Antitrust Ruling Chills Standards Setting: A U.S. Supreme Court Ruling Has Clouded Standards Setting with the Threat of Antitrust Liability, *IEEE Spectrum* 19(8): 52–54.

Pressman, David, and Richard Stim, 2006, *Nolo's Patents for Beginners*, Berkeley, CA, Nolo.

Purcell, Donald E., and William E. Fox Jr., 2001, Associations, Safety Standards and Liability, *SES Standards Engineering* 53(3): 1–5.

Ruger, Alison C., 2001, Standards Dissemination and Digital Rights Management: A Reseller's Point of View, *SES Standards Engineering* 53(4):1–4.

Ruggieri, James Angelo, 2004, Ipse Dixit (Something Asserted But Not Proved) and the Use of Standards in Forensics, *SES Standards Engineering* 56(4): 1–7.

Samuelson, Pamela, 2006, Copyrighting Standards: Should Standards Be Eligible for Copyright Protection? *Communications of the ACM* 49(6): 27–31.

Sherr, Sava L., 1982, Societal Aspects of Engineering Standards, *IEEE Technology and Society Magazine* 1(3): 13–16.

Shurmer, Mark, and Gary Lea, 1995, Telecommunications Standardization and Intellectual Property Rights: A Fundamental Dilemma? *ACM StandardView* 3(2): 50–59.

Smoot, Oliver, 1995, Tension and Synergism between Standards and Intellectual Property, *ACM StandardView* 3(2): 60–67.

Swankin, David A., 1990, How Due Process in the Development of Voluntary Standards Can Reduce the Risk of Anti-Trust Liability, NIST/GCR 90/571.

Thorpe, James F., and William H. Middendorf, 1979, *What Every Engineer Should Know About Product Liability*, New York, Marcel Dekker.

Traver, A. E., March 29–April 1, 1982, The Effect of Product and Safety Standards on Litigation, Contributed by the Design Engineering Division of The American Society of Mechanical Engineers for Presentation at the Design Engineering Conference, Chicago.

Twomey, David P., Marianne Moody Jennings, and Ivan Fox, 2002, *Anderson's Business Law And the Legal Environment*, 18th ed., Cincinnati, Ohio, West-Thomson Learning, pp. 474 et seq.

Updegrove, Andrew, 1995, Consortia and the Role of the Government in Standard Setting, in *Standards Policy for Information Infrastructure*, edited by Brian Kahn and Janet Abbate, Cambridge, MA, The MIT Press, p. 345.

U.S. Department of Commerce, 1979, U. S. Department of Commerce Model Uniform Product Liability Act (MUPLA), *Federal Register* 44(212): 62714–62750.

Veeck v. SBCCI, 1999, In the United States Court of Appeals for the Fifth Circuit (No. 99-40632), Peter Veeck, doing business as Regional Web, Plaintiff-Counter Defendant-Appellant versus Southern Building Code Congress International, Inc. Appeals from the United States District Court for the Eastern District of Texas, February 2, 2001.

Warren-Boulton, Frederick R., Kenneth C. Baseman, and Glenn A Woroch, 1995, Economics of Intellectual Property Protection for Software: The Proper Role for Copyright, *ACM StandardView* 3(2): 68–78.

11

Federal Roles in Standards, Conformity Assessment, and Accreditation

Background

The current partnership of the federal government and the standards and conformity assessment community did not always exist. In fact, as Howard I. Forman (1981) pointed out, the battered and bruised standards community was rather wary of federal involvement in standards after the *Hydrolevel* and *Allied Tube* cases and the FTC Notice of Proposed Rulemaking on standards and certification. Other authors also cast a jaundiced eye on the federal role in standards making in the information and communications technology (ICT) field (Aiken and Cavallini 1995). Linda Garcia (1993) presented a well-researched paper advocating that the government assume a more significant role than in the past. In light of the growing importance of standards and conformity assessment in trade, the National Institute of Standards and Technology (NIST) held hearings on April 3 to April 5, 1990, to explore ways to improve U.S. participation in international standards-related activities *and to identify possible government actions* (Leight 1990). A total of 65 individuals and representatives of organizations appeared before an expert panel and written comments were received from 257 individuals and organizations. It was during this process that a potential model for the United States, patterned after the Standards Council of Canada, a model that became known as Standards Council of the United States (SCUSA), was aired. Nearly half of the commenters felt that government must play a major role. A seminal paper published by the National Standards Policy Advisory Committee (NSPAC) posited that cooperation between government and the private standards community is an essential element. "The national interest is best served when both the government and private sectors, and/or components of both sectors, initiate, develop and participate in programs which stimulate and encourage cooperation of both sectors in national standards activities" (NSPAC 1978). Most of the NSPAC was not adopted but, fortunately, that attitude prevailed in the long term. In 1998, NIST and ANSI cohosted a summit conference, Toward a National Standards Strategy to Meet Global Needs (Leight and Leuteritz 1998).

Overview of Federal Government Activities in Conformity Assessment

An overview of the conformity assessment and related programs of eighteen agencies of the federal government is presented in NIST Special Publication 739 (Breitenberg 1999). The information includes the products, agency contacts, the date the program was initiated, the form of compliance (voluntary or mandatory), the legal authority (Code of Federal Regulations), the aim of the program and its benefits, the methodology (including testing and inspection), the conformity identification, the availability of documents, the obligations of the manufacturer/vendor, the enforcement methods, the terms (periods) that certifications are valid, any reciprocity, the defining standards, codes and regulations, and keywords.

The agencies covered include the Departments of Agriculture, Commerce, Defense, Energy, Health and Human Services, Housing and Urban Development, Interior, Justice, Labor, Transportation, Treasury, Veterans Affairs; and the independent agencies of the Consumer Product Safety Commission, Environmental Protection Agency, Federal Communications Commission, Federal Trade Commission, General Services Administration, and the Nuclear Regulatory Commission.

The federal territory outlined is far too complex to be covered here in a meaningful way so a sample of significant federal activities is described in the following.

Administrative Procedures Act (5 U.S.C., 1976, as Amended)

The Administrative Procedures Act (APA), as implemented by the Administrative Conference of the United States, directly impinges on standards and conformity in at least two ways. The first is the way government agencies make rules. A *rule* is defined in Black's Law Dictionary as "an established *standard*, guide or regulation" (emphasis added). The second way the APA impinges on standards involves the authority for the use of agency "incorporation by reference" of standards and similar documents in its regulations.

Section 553 of Title 5, United States Code, is the APA's general rulemaking section, and rulemaking governed by it is commonly called "informal" or "notice and comment" rulemaking. A generic outline of the process follows:

1. *Agency* announces proposed *rule* in the *Federal Register* for public review and comment (NPRM).

2. A ninety-day period is provided for review and for submission of comments to *Agency* on proposed *rule*.

3. At the same time that the proposed *rule* is announced in the *Federal Register*, *Agency* sends the proposed *rule* to senior contacts in federal government agencies and departments and to state agencies.

4. Comments received in response to the *Federal Register* Notice and other notifications are reviewed by *Agency* to determine if modifications to the proposed *rule* are needed.

5. A detailed justification document is prepared by *Agency*, analyzing the comments received and explaining why modifications (if any) are made, or explaining why recommended changes were *not* made.

6. *Agency* submits recommendation on adoption to the Department (of which it is a part, if any), all the while being subjected to various political pressures.

7. A notice announcing approval of the *rule* is published in the *Federal Register*.

8. Housekeeping and related actions and notifications are then carried out.

This process can be very complicated and lengthy when complex technical issues are involved and that tends to make federal rulemaking very slow and uncertain in such circumstances. Diane Thompson (1991) explained the many advantages to the government, industry, and consumers of the use of voluntary standards in regulations.

The "boilerplate" citation in the Code of Federal Regulations (CFR) when an agency incorporates by reference reads something like "Material Approved for Incorporation by Reference: The Director of the Federal Register has approved under 5 U.S.C. 552(a) and CFR Part 51 the incorporation by reference of the following publications." Approximately 5,000 Standards have been incorporated by reference in federal regulations, since the passage of the National Technology Transfer and Advancement Act (NTTAA), with approximately 600 from ASTMI alone. That does not count all of the uses of voluntary standards by the Department of Defense over many years, which brings the total to more than 9,000.

OMB Circular A-119 (First Take)

The Office of Management and Budget (OMB) is lodged in the Office of the President. OMB prepares the president's Federal Budget and performs many management functions in directing and coordinating various federal executive agencies. OMB issues various guidance documents to federal agencies, including *Circulars*.

On January 17, 1980, OMB issued Circular A-119: Federal Participation in the Development and Use of Voluntary Standards. It had long been recognized that special government standards, such as the MIL-STD-806B for digital logic diagrams, were not only wasteful and expensive to prepare but greatly increased the transaction costs when a government agency used such standards in procurement. MIL-STD-806B was superceded by ANSI Y32.14-1973 (an IEEE standard), which used rectangular logic symbols that were aligned with those of a corresponding IEC standard. The ANSI standard was adopted by the Department of Defense. OMB Circular A-119 directed executive branch agencies to work with standards developers and also to adopt such standards in procurement. A primary responsibility for implementing these policies was vested in the Secretary of Commerce. The circular provided for the identification of Standards Developing Organization (SDOs) that could participate in the program as those using "due process" in their operation. OMB A-119 also established the Interagency Committee on Standardization Policy (ICSP) in the Commerce Department. In addition to proposing a (commerce) department-sponsored voluntary dispute resolution service that would address procedural complaints, OMB A-119 proposed that annual reports to OMB on progress toward the goals of adopting voluntary consensus standards be made.

On June 2, 1980, the Department of Commerce issued a notice: "Implementation of Federal Voluntary Standards Policy; Proposed Procedures for Listing Voluntary Standards Bodies Eligible for Federal Agency Support and Participation, and for a Department Sponsored Voluntary Dispute Resolution Service for Procedural Complaints Against Listed Voluntary Standards Bodies." The department contacts for further information were Dr. Howard I. Forman, Deputy Assistant Secretary for Product Standards Policy, and Mr. Donald R. Mackay. On April 21, 1982, OMB published a draft circular for comments. The final issuance of Circular A-119 did not include the mandatory procedural requirements and made additional changes; it was published on November 1, 1982 (OMB 1982). Circular A-119 was revised in October 1983 and again in October 1993. More on OMB A-119 appears in a later discussion of the National Technology Transfer and Advancement Act of 1995.

The Department of Defense (DOD)

As indicated earlier, the Department of Defense has long been a leader in the use of "voluntary consensus standards" (VCSs, but "Standards" herein). Many cost savings in the so-called MILSPEC REFORM have been documented (DOD 1996). The goals of DOD acquisition reform and the ways it is being carried out are documented by Gregory E. Saunders and Stephen C. Lowell in a paper published in 1995.

National Institute of Standards and Technology (NIST)

The National Bureau of Standards (NBS) formed in 1901 is the "national metrology laboratory," entrusted with the establishment and maintenance of the *physical* standards (including weights and measures) on which U.S. commerce and science depend. The weights and measures laboratories of the states have always traced their calibrations to NBS. NBS was renamed NIST and expanded in 1988 pursuant to the Omnibus Trade and Competitiveness Act, which tasked NIST with many of its present duties related to trade. Oppermann (2005) has reviewed the weights and measures role played by NIST. An overall survey of NIST activities appears in *Standards Engineering* (SES 2005). NIST produces many documents on standards and provides many official functions for the U.S. standards community including the operation of the National Center for Standards and Certification Information (NCSCI), which gives early warning on proposed foreign regulations, a function required by the WTO TBT Agreement. NIST also administers the National Technology Transfer and Advancement Act as described later. NIST is also the custodian of FIPS, the Federal Information Processing Standards (Keblawi and Sullivan 2007).

National Voluntary Laboratory Accreditation Program (NVLAP) and the National Voluntary Conformity Assessment System Evaluation (NVCASE) Program

The original version of the NVLAP was briefly described in Chapter 8 as a potential national accreditor of testing laboratories. At the time of its commencement in 1976, there was no national accreditor of testing laboratories, a badly needed function. The program was started by Howard I. Forman who also was very instrumental in the formation of the International Laboratory Accreditation Conference (ILAC). The birth of ILAC was documented in a three-part paper by Forman that appeared in *Standards Engineering* (Forman 1989a, 1989b, 1989c). ILAC accreditation procedures formed the basis of the ISO Guide 25 (later ISO/IEC Guide 25). Many Conformity Assessment Bodies, including Underwriters Laboratories, sought to obtain Laboratory

Accreditation Programs (LAPs) for their areas of interest. Early LAPs were established for thermal insulation materials, freshly mixed field concrete, and carpet. The laboratories doing EMC measurements needed an accreditation that was acceptable to the FCC. Other laboratories needed programs for other technical areas including MIL-STD-462 (Electromagnetic Compatibility) Test Methods. Calibration services were added to NVLAP in 1992 and many clients were accredited in an Ionizing Radiation Dosimetry Program. An LAP was established for an Airborne Asbestos Analysis Program to support the EPA. By 1994, NVLAP fields of accreditation included acoustics, energy efficiency, thermal insulation, and several others. The NVLAP Procedures and General Requirements were collected in NIST Handbook 150 in March 1994. The handbook included the requirements of ISO/IEC Guide 25 (which later became ISO/IEC International Standard 17025). As of 1997, a total of 676 laboratories were accredited. The current NVLAP programs are described in Title 15—Commerce and Foreign Trade, chapter II—National Institute of Standards and Technology, Department of Commerce, Part 285.

The NVLAP program was, in part, a model for the National Voluntary Conformity Assessment Systems Evaluation (NVCASE) Program, also in NIST. In March of 1992, NIST published a "Request for Comments on a Proposal to Establish the Conformity Assessment Systems Evaluation Program (CASE)." The "Background of the Request for Comments" noted that the need for such a program followed a joint communiqué issued by then Commerce Secretary Mosbacher and European Community (EC) Commission Vice President Bangemann in June 1991. The EC and the governments of other trading partners indicated a desire to deal with a government entity that could provide assurance of the validity of U.S. conformity assessment activities pertaining to products *regulated* in foreign countries (emphasis added; i.e., the program does not apply to unregulated products). The comments on the request are summarized in NISTIR 5138 (Donaldson and Cooke 1993). NIST issued a Notice of Proposed Rule to establish the NVCASE Program in April 1993 (Warshaw 1993). The program was announced on April 22, 1994, and is documented in 15 CFR Subtitle B, Ch. II Part 286. The *Scope* is defined in Section 286.2. The program activities are classified by level:

1. *Conformity level*: This level encompasses comparing a product, process, service, or system with a standard or specification. As appropriate, the evaluating body can be a testing laboratory, product certifier or certification body, or quality system registrar.

2. *Accreditation level*: This level encompasses the evaluation of a testing laboratory, a certification body, or a quality system registrar by an independent body—an accreditation body—based on requirements for the acceptance of those bodies, and granting of accreditation to those that meet the established requirements.

3. *Recognition level*: This level encompasses the evaluation of an accreditation body based on requirements for its acceptance, and the recognition by the evaluating body of the accreditation body that satisfies the established requirements.

Section 286.3 recited the **Objective**: The objective of the program is to identify the activities of requesting U.S.-based conformity assessment bodies that have been evaluated as meeting requirements established for their acceptance by foreign governments. The evaluation may be provided by NIST or by bodies recognized by NIST for this purpose under the scope of this program. Like the NVLAP Program, NVCASE has its generic and special requirements.

As a result of the successful initiation of the NVCASE Program, the U.S.–EU Mutual Recognition Agreement was reached that identified U.S. EMC test laboratories and U.S. radio and telecommunication terminal equipment test laboratories in December 2003 and January 2005, respectively.

Fastener Quality Act (P.L. 101-592, as Amended Under P.L. 104-113, P.L. 105-234 and P.L. 106-34)

In July 1988, a Committee Print of the 100th Congress, 2nd Session, publicized a major horror story about how defective fasteners (nuts, bolts, etc.) had permeated both government and commercial activities in the United States. The report was titled "The Threat from Substandard Fasteners: Is America Losing Its Grip?" (U.S. Congress 1988). The report summarized an eighteen-month investigation, five days of open hearings, and two sessions that remain closed. The report "documents the frightening variety of end uses where substandard nuts, bolts, screws, and washers *threaten the safety of every American*" (emphasis added). The report includes a litany of examples where defective fasteners resulted in "deaths that have reportedly resulted from industrial and aircraft accidents because of fastener failure." The report goes on to cite the "inadequacy of a system grounded on voluntary consensus standards, which are technically sound but not adequately enforced, to provide the quality assurance and traceability needed in today's society." The issue here was not standards related, as there were suitable Society of Automotive Engineers (SAE), American Society for Testing and Materials (ASTM), and American Society of Mechanical Engineers (ASME) standards, but one of *conformity assessment*. These were only two of thirteen points summarized in the summary and introduction. The findings section of the report cited numerous cases where high-performance fasteners (e.g., SAE grade 8.0) were mismarked and counterfeit, with numerous examples of suppliers, some domestic but mostly foreign, deliberately supplying fasteners known to be defective! It is interesting that when procurement officials tried to eliminate some of these suppliers, the Small Business Administration (SBA) came to the defense of the miscreants. The report also identified fifty companies that had supplied defective or bogus fasteners. The report includes a number of recommendations including the enactment of legislation that would "require manufacturers and importers to submit all lots of fasteners with significant safety applications to accredited laboratories for testing to assure they meet the physical, chemical, and dimensional specifications corresponding to the terms of their sale." On November 16, 1990, HR 3000 became law as the Fastener Quality Act (P.L. 101-592).

The following is a textbook example of the Administrative Procedures Act described earlier (with some details left out for brevity). A Regulatory Impact Analysis/Regulatory Flexibility Analysis for Fastener Quality Act (FQA) was issued on August 5, 1992, but had to be revised later to take into account the changes mandated by the 1996 NTTAA. The FQA directed the Secretary of Commerce to apply NVLAP to the Fastener Act issues. On Monday, August 17, 1992, the Department of Commerce, through the National Institute of Standards and Technology, issued a Notice of Proposed Rulemaking on Procedures for Implementation of the Fastener Quality Act: Proposed Rule. Meanwhile, ASME announced its own Fastener Accreditation Program. ASTM Committee F-16 on Fasteners held a symposium in November 1994 to explore ways to comply with the FQA. A Fastener Advisory Committee and a Public Law Task Force (in 1995) were consulted by NIST for the making of a final rule. Meanwhile, a new subsection (5(d) of the proposed Rule) was modified by Public Law 104-113, The National Technology Transfer and Advancement Act (NTTAA) was signed into law on March 7, 1996. Revisions were developed in light of the "legislative intent" of Congress in the FQA. Finally, the Department of Commerce issued a *Final Rule* on September 26, 1996, along with an analysis of the comments and other inputs (Department of Commerce 1996). An interesting "Note: The 'Final Rule'" of the FQA listed the DOC director's list of "Consensus Standards Organizations" as had been proposed by Howard I. Forman eighteen years earlier (see earlier discussion on NSPAC). P.L. 105-234 exempted certain aircraft fasteners from the act.

As the old saying goes, it isn't over 'til the fat lady sings. The fat lady in this case included representatives from the automobile industry, the Transatlantic Business Dialog, and an industry group led by the Industrial Fasteners Institute. These groups, to the concern of some, persuaded Congress to replace the original FQA with a "replacement Fastener Quality Act" that was signed into law December 6, 1999 (P.L. 106-34). The "Fastener Quality Act Amendment Act" (sometimes called FQA99 for brevity) basically gutted the original FQA, as amended, and replaced it because "state-of-the-art manufacturing and improved quality assurance systems have dramatically improved fastener quality, so virtually all fasteners sold in commerce meet or exceed the consensus standards for the uses to which they are applied."

Consensus standards organizations were identified and defined in the new FQA and the supply of fasteners was put back under the control of the private sector quality management standards, such as the ISO 9000 series, and derivative standards and others. The exemption of certain "unhardened fasteners" covered by several ASTM standards as well as most aircraft fasteners was also included in FQA99. The detection of bogus fasteners was turned over to a "hotline" operated by the Department of Commerce, which could certify to the Justice Department cases of violations of the act (which still carries both civil and criminal penalties for violations). A "List of Self-Declared Accreditation Bodies" was also authorized by the act, even though the NVLAP list of accredited bodies was discontinued.

The legislation resulting in FQA99 was opposed by some, including the National Society of Professional Engineers (NSPE). There continued to be public reports of fastener problems, including aircraft fasteners, well into 2003, including an allegation that defective bolts had endangered the Shuttle Columbia (*Austin American-Statesman* 2003). The bogus fastener problem was only a mountain, although a major peak, in a mountain range of bogus products ranging from handbags to semiconductors and prescription drugs with which the United States is still coping (Balfour 2005).

National Technology Transfer and Advancement Act (NTTAA) of 1995 (P.L. 104-113), and The OMB Circular No. A-119 (Revised February 10, 1998)(Take 2)

The National Technology Transfer and Advancement Act (NTTAA) of 1995 was signed into law on March 7, 1996; it was "An Act to Amend the Stevenson-Wydler Technology Innovation Act of 1980" with respect to inventions made under cooperative research and development agreements, and for other purposes. The major thrust of interest here is in Section 12, Standards and Conformity. The act required NIST "to compare standards used in scientific investigations, engineering, manufacturing, commerce, industry, and educational institutions with the standards adopted or recognized by the Federal Government and to coordinate the use by Federal agencies of private sector standards, emphasizing where possible, the use of standards developed by private consensus organizations" and assigned the following task to NIST: "to coordinate Federal, State and local technical standards activities and conformity assessment activities, with private sector technical standards activities and conformity assessment activities, with the goal of eliminating unnecessary duplication and complexity in the development and promulgation of conformity assessment requirements and measures."

The NTTAA directs (with some exceptions) all federal agencies and departments to use technical standards that are developed or adopted by voluntary consensus standards bodies, using such technical standards as a means to carry out policy objectives or activities determined by the agencies and departments. The NTTAA also directs federal agencies and departments to consult with voluntary, private sector, consensus standards bodies, and, when appropriate, participate with such bodies in the development of technical standards (Howard I. Forman's vision has been achieved!). JoAnne Overman (2006) reviewed the NTTAA and its first decade of use.

The National Institute of Standards and Technology, through its management of the Interagency Committee on Standards Policy (ICSP) and otherwise, thus becomes a major player in the private and public sectors of the standards community. The NTTAA codified the 1993 Revision of the OMB Circular A-119. In light of the increased federal participation in private standards development, ANSI and ASTM have developed guides for use by their SDOs and committees when eventual federal use of standards is anticipated. At an ANSI Panel Meeting in 1997, Howard Seltzer stated, "One of the more attractive aspects of the NTTAA would be that agencies could—with a standard that has had meaningful consumer participation in its development—go to a direct final rule, publish it and allow 30 days for comment. Unless there was significant negative comment, it would become a rule."

The OMB Circular A-119 was revised February 10, 1998, to better coordinate it with the NTTAA. On Thursday, February 19, 1998, the Executive Office of the President, OMB, reported the results of a December 27, 1996, Notice and Request for Comments on Proposed Revision of OMB Circular A-119 (61 FR 68312). The comments were reviewed and the Final Revision of the Circular issued. The Revised OMB Circular A-119 was designed to better coordinate it with the NTTAA. It provides definitions and guidance to agencies in making their reports to OMB. These yearly reports were required to show each special government unique standard that was developed rather than the adoption of a suitable private standard. These agency reports are made to OMB via NIST. The revised OMB Circular provides definitions and guidance in the form of question-and-answer format. The circular also directs the Secretary of Commerce to issue policy guidance for conformity assessment. Kathleen Kono (1998) describes the Revised OMB Circular A-119. Kono's paper also has a summary of the number of ASTM members in various government agencies and the number of ASTM standards referenced in the CFR by each agency. Paul Gill et al. (2000) have provided a summary of federal agency participation in Standards Bodies.

Conformity Assessment under the NTTAA

Section 12(b) of the NTTAA modified existing 15 U.S.C. 272; Part 287—Guidance on Federal Conformity Assessment, which is reviewed here. Part 287.1(a) states in pertinent part: "This part provides guidance for each Federal agency to use in evaluating the efficacy and efficiency of its conformity assessment activities. Each agency should coordinate its conformity assessment activities with those of other appropriate government agencies *and with those of the private sector to reduce unnecessary duplication* (emphasis added)." Part 287 includes definitions of *Accreditation, certification, conformity assessment, inspection, supplier's declaration of conformity, registration and testing based on ISO/IEC Guide 2 (1996).* Part 287 also defines *Agency Standards Executive* as an official designated by an agency as its representative on the Interagency Committee for Standards Policy (ICSP) and delegated the responsibility for agency implementation of OMB Circular A-119 and the guidance in this part. *Recognition* is defined as "a procedure used to provide formal notice that an accreditation body is competent to carry out specific tasks. These tasks include: the accreditation of testing laboratories and inspection, certification and registration bodies. A governmental recognition system is a set of one or more procedures used by a Federal agency to provide recognition." It should be noted that "Recognition" as used in Part 287 is not the same as the "Recognition" process of the Occupational Safety and Health Administration (OSHA) in the Nationally Recognized Testing Laboratories described in Chapter 2.

Additional parts of Part 287 define the responsibilities of the National Institute of Standards and Technology in the management of the ICSP process. The responsibilities of federal agencies are also outlined (i.e., to implement the policies contained in Part 287). In addition to providing a rationale for its use of specified conformity assessment procedures and processes in rulemaking and procurement actions to the extent feasible, Part 287 also encourages the use of relevant guides or standards for conformity assessment practices published by domestic and international standardizing bodies as appropriate in meeting regulatory and procurement objectives, including those of ANSI, ISO, IEC, ITU, and others. Agencies are encouraged to participate in efforts designed to improve coordination among governmental and private sector conformity assessment activities including those of the National Cooperation for Laboratory Accreditation (NACLA), the National Environmental Laboratory Accreditation (NELAC), the ISO/CASCO, ANSI, and ICSP working groups dealing with conformity assessment. Agencies are also encouraged to participate in the development of private sector conformity assessment standards to ensure that federal viewpoints are represented. Responsibilities of Agency Standards Executives, in addition to those described in OMB Circular A-119, are also outlined.

One of the most important functions of NIST for the standards and conformity community is the Technology Services, which has produced many of the readily available sources of information on standards and conformity assessment. The guides listed in Appendix 5 are only a fraction of that material. Pursuant to the NTTAA, NIST has developed a new standards Web site: Standards.gov.

Federal Communications Commission (FCC)

The Federal Communications Commission (FCC) was formed by the Communications Act of 1934 when it became apparent that radio frequency interference (RFI) was becoming a very serious problem that threatened radio communications of all kinds. Thus an important function of the FCC is spectrum management, which involves the International Telecommunications Union (ITU) as well as national issues (see Chapter 6). The classical method of spectrum management is the assignment of radio transmitters to specified frequencies by federal regulation and control of radiated power and coverage. Radios were good radiators of local oscillator signals so technical rules (standards) were developed to address that and similar problems.

With the advent of television, receiver (and transmitter) standards became crucial for interoperability in the TV system of broadcasting. The Radio Manufacturing Association (RMA) developed the basic television standard (Fink 1940). That standard was later adopted by the National Television Standards Committee (NTSC) under the aegis of the FCC, the major change being the increase of the number of scanning lines from 441 to 525. Standards making was relatively simple in those days (before the Administrative Procedures Act).

Development of the High-Definition Television (HDTV) Standards

Fast forward to the modern era where government agencies and the FCC presided over *and participated in* the development of the HDTV standards. In a study titled "The Government's Role in the HDTV Standards Process: Model or Aberration?" HDTV efforts in Europe and Japan were aimed at refining the old analog process, efforts that were doomed to fail (Neil et al., 1995). In the United States, there developed a unique government/industry/academia team, which was a new concept in standards development. Several government agencies, including the Department of Commerce, the FCC Advisory Committee on Advanced Television Systems (ACATS), and the Advanced Research Projects Agency (ARPA) worked with disparate industry cultures (computers and TV) to develop, against all odds, a totally new digital concept for HDTV. The process became one involving the important national economics as much as complex technology standards. Thus the technical basis for digital TV was the product of industry and several federal agencies, including the Department of Defense, working under the continuing guidance of the FCC to develop a digital standard for HDTV. The question in the title was answered in favor of a *model*, although a complex one, of how the government has become an actor, or participant, in the standards development process. The news is not all good, however. One of the primary companies that made major contributions to the process was the Zenith Electronics Corp. (Bretl et al., 2006). That paper and others trace the development of global digital television in a special issue of the Proceedings of the IEEE (January 2006). Zenith was destroyed in part because the process took so long and Zenith was forced into bankruptcy by the process (among other things) and was bought up as a "design lab" by a South Korean electronics firm, LG Electronics. Thus, the last American-owned TV manufacturer was demolished in the lengthy process.

FCC Measurement Standard for Computer Emissions (47 CFR 15.31, 1993)

In the late 1970s, the radio frequency emissions from computers produced significant interference with emergency and other radio services. As a result, the FCC commenced to regulate those emissions. As computer technology progressed, so did the FCC measurement standards required for implementing the regulations. In the 10-1-93 edition of 47 CFR Part 15, the FCC mandated that the original measurement standard had to be replaced. The Administrative Procedures Act was initially followed for the development of the new measurement standard. People came out of the woodwork to oppose and modify the proposed changes. The FCC soon realized that the APA process was not working very well in this instance so it punted the issue to ANSI C63.4 committee (an ANSI-recognized IEEE committee). The development of a new measurement standard was slowed by the rapid turnover of the committee members (Edwin L. Bronaugh, personal communication 2007), but the C63.4 finally produced a suitable measurement standard that was adopted, with several exceptions, by the FCC. This scenario pointed up the inherent difficulty of using the APA to develop complex technical standards.

FCC Adopts Technically Superior ATIS Standard

The FCC had proposed a standard for "911 only wireless phones," a so-called consecutive number approach. The Alliance for Telecommunications Industry Solutions (ATIS), via its Emergency Services Interconnection Forum (ESIF) proposed an alternative that was a "more far-reaching and technically superior solution" that was adopted by the FCC (ATIS 2003). The ATIS standard permitted the Public Safety Answering Point (PSAP) to distinguish legitimate 911 calls from abusive nonemergency calls.

In summary, the examples described here show that in many cases the standards development process of the Administrative Procedures Act is less than satisfactory and points up the importance of the NTTAA partnership of government and private sector standards development. One retiring official of the FCC remarked that the development of the Modular Plug and Jack Standard (47 CFR Part 68) was one of the best standards produced by the FCC.

References

Aiken, Robert J., and John S. Cavallini, 1995, When Are Standards Too Much of a Good Thing? Will They Provide Interoperability for the National Information Infrastructure? In *Standards Policy for Information Infrastructure,* edited by Brian Kahin and Janet Abbate, Cambridge, MA, MIT Press, part 4.

Alliance for Telecommunications Industry Solutions (ATIS), 2003, FCC Finds ATIS Standard Technically Superior, *SES Standards Engineering* 55(6): 7–8.

Austin American-Statesman, 2003, Bolt may threaten shuttles, June 13, p. A2.

Balfour, Frederick, 2005, FAKES! The Global Counterfeit Business Is Out of Control, Targeting Everything from Computer Chips to Life-Saving Medicines. It's Getting So Bad That Even China May Need to Crack Down, *Business Week,* February 7, pp. 54–64.

Breitenberg, Maureen, ed., 1999, *Directory of Federal Government Certification and Related Programs,* Gaithersburg, MD, U.S. Department of Commerce, Technology Administration, National Institute of Standards and Technology, NIST Special Publication 739.

Bretl, Wayne, William R. Meintel, Gary Sgrignoli, Xianbin Wang, S. Merrill Weiss, and Khalil Salehian, 2006, ATSC RF, Modulation, and Transmission, *Proceedings of the IEEE* 94(1): 44–59.

Department of Commerce (DOC), 1996, National Institute of Standards and Technology, 15 CFR Part 280, Procedures for Implementation of the Fastener Quality Act, Final Rule, *Federal Register* 61(188): 50538 et seq.

Department of Defense (DOD), 1996, MILSpec Reform: Results of the First Two Years, *SES Standards Engineering* 48(5): 1–5.

Donaldson, John L., and Patrick W. Cooke, 1993, A Program for Conformity Assessment System Evaluation: Analysis of Comments on the NIST Proposal (NISTIR 5138), Gaithersburg, MD, U.S. Dept. of Commerce, Technology Administration, National Institute of Standards and Technology.

Fink, Donald G., 1940, *Principles of Television Engineering,* New York, McGraw-Hill.

Forman, Howard I., 1981, The Voluntary Standards System: Government Intervention, Government Cooperation, or Laissez-Faire? Presentation at the 1981 Standards Engineering Society Conference of 1981.

Forman, Howard I., 1989a, General Description of the International Laboratory Accreditation Conference (ILAC) and Its Relation with Standardization and Quality Certification, *SES Standards Engineering* 41(1): 12–22.

Forman, Howard I., 1989b, General Description of the International Laboratory Accreditation Conference (ILAC) and Its Relation with Standardization and Quality Certification, *SES Standards Engineering* 41(2): 37–42.

Forman, Howard I., 1989c, General Description of the International Laboratory Accreditation Conference (ILAC) and Its Relation with Standardization and Quality Certification, *SES Standards Engineering* 41(3): 55–68.

Garcia, Linda, 1993, A New Role for Government in Standard Setting? *ACM StandardView* 1(2): 2–10.

Gill, Paul, William H. Vaughan, and Stephen Lowell, 2000, Participation by Federal Agencies in Voluntary Consensus Standards Bodies, *SES Standards Engineering* 52(3): 1–7.

Keblawi, Feisal, and Dick Sullivan, 2007, The Case for Flexible NIST Security Standards, *IEEE Computer* 40(6): 19–26.

Kono, Kathleen, 1998, OMB A-119 Revised, A Tool to Implement the National Technology Transfer and Advancement Act, *ASTM Standardization News* 26(6): 19–21.

Leight, Walter G., 1990, Government's Role in Standards-Related Activities: Analysis of Comments, NISTIR 4367, Gaithersburg, MD, U.S. Department of Commerce, National Institute of Standards and Technology, Office of Standards Services.

Leight, Walter G., and Krista J. Johnsen Leuteritz, 1998, Toward a National Standards Strategy, Conference Summary Report (NISTIR 6259), Gaithersburg, MD, U. S. Department of Commerce, Technology Administration, National Institute of Standards and Technology.

National Standards Policy Advisory Committee (NSPAC), 1978, *National Policy on Standards for the United States and a Recommended Implementation Plan*, Gaithersburg, MD, American National Standards Institute.

Neil, Suzanne, Lee McNight, and Joseph Bailey, 1995. The Government's Role in the HDTV Standards Process: Model or Aberration?, in *Standards Policy for Information Infrastructure*, edited by Brain Kahin and Janet Abbate, Cambridge, MA, The MIT Press, pp. 276–288.

Office of Management and Budget (OMB), 1982, Issuance of Circular No. A-119, Federal Participation in the Development and Use of Voluntary Standards, *Federal Register* 47(211): 49496–49499.

Oppermann, Henry, 2005, NIST's Role in Weights and Measures, *ASTM Standardization News* 33(1): 34–37.

Overman, JoAnne R., 2006, The NTTAA: 10 Years of Public-Private Partnership, *SES Standards Engineering* 58(1): 1–8.

Saunders, Gregory E., and Stephen C. Lowell, Department of Defense Acquisition Reform: A New Way of Doing Business, *ASTM Standardization News* 23(9): 46–53.

Seltzer, Howard, 1997, Panel IV: Consumer Involvement in the Standards Development Process, *ANSI Reporter* (Special supplement), pp. 10–12.

Standards Engineering Society (SES), 2005, Focus On … NIST, National Institute of Standards and Technology, *SES Standards Engineering* 57(1): 16–17.

Thompson, Diane, 1991, The Benefits of Using Voluntary Standards in the Legislative and Regulatory Processes, *ASTM Standardization News* 19(1): 18–19.

U.S. Congress, House, July 1998, The Threat from Substandard Fasteners: Is America Losing Its Grip? A Report on the Subcommittee on Oversight and Investigations of the Committee on Energy and Commerce, Committee Print 100-Y.

Warshaw, Stanley, 1993, National Voluntary Conformity Assessment Systems Evaluation (NVCASE) Program of NIST, NPRM, *Federal Register* 58(78): 24146 et seq.

12

Management Aspects of Standards, Conformity Assessment, and Accreditation

Two Cultures

One writer on the subject of management and standards indicated that managers should learn more about standards and standards people should learn more about management! That there has been a gulf separating these groups is fairly clear. The reader is warned that general statements regarding the issues in this chapter may not be applicable to both the ICT (information and communication technology) and non-ICT areas of standardization. Moreover, general statements have a habit of being invalidated by time as conditions and technologies change. David Hemenway (1975), in his classic work on standards, points out that managers think in terms of *money*, whereas standards people tend to think in terms of *things* and *processes*. Hemenway was one of the first writers to seriously address the economic aspects of standards; he showed quantitatively in appendix A of his book how product standards produce economy of scale benefits. That proposition seems intuitively obvious but it had not often been treated analytically. The Standards Engineering Society (SES) addressed the economic aspects of standards in several of its annual meetings. Some of those results were summarized in a pioneering SES publication *The Economics of Standardization* edited by Robert Toth (1984). The single product standard, now known as a form of *variety reduction*, has been a cornerstone of the economic principles of standardization. Toth's book, *Standards Management: A Handbook for Profits* (1990), is a tour de force on the many ways that standards and certification can contribute to the bottom line of companies. Although a little dated, many of the papers in the compendium can still be mined for useful information on a wide variety of subjects involving management and standards. A variety of authors describe how company standards groups can be justified and organized in various environments. Toth's book is definitely a how-to treatment for standardization at the company level and is a must-have reference for both managers and standards people.

Economics (Benefits and Costs) of Standardization

Most of the earlier references on the economics of standardization addressed what might be called the *microeconomics* issues of the subject, that is, the issues were addressed at company levels as opposed to national or global levels, which are the domain of *macroeconomics* (Frank and Bernanke 2004). Clifford Lynch has published a critique of the present economics of standards development and distribution (Lynch 1999). Lynch posits that one of the factors in the rise of consortia, and the decline of SDOs, is the high cost of participation and access to standards. Economist Joseph Farrell (1990) has published a seminal paper for "noneconomists" explaining some economic-standards terms including compatibility, lock-in, switching costs, and collusion among vendors. Many of the published papers in this area appear in the *Rand Journal of Economics,* successor to the *Bell System Journal of Economics.* The relationship of dominant designs and standards is studied by Scott Gallagher (2007) who clarifies the differences between these two concepts.

Microeconomics of Standardization

Company standards that address compliance with Regulations do not generally have to be justified, but the efficiency with which such standards operations implement the company compliance efforts is fair game for management. Managers should be sufficiently enlightened that they understand the rationale for the Regulations and the need to comply with them. The subculture of Regulatory standards and conformity assessment is often called Regulatory Compliance Engineering, a subject beyond the scope of the present work.

In Chapter 18 of *Standards Management,* Toth (1990) summarizes the benefits, both tangible and intangible, of standardization (Figure 18-1). Toth also summarizes the tangible and intangible costs and disadvantages of standardization in Figure 18-2. Toth then qualitatively reviews the Revenues (benefits) as compared with the costs of standardization. The way in which the use of a company standard can benefit the company, especially through variety reduction, is illustrated graphically in Figure 18-7.

Among the interesting papers in *Standards Management* is one by James C. Miller (1990) on how a company standards program can be evaluated, an emergent form of the "standards strategies" discussed later.

Part A of a recent report by Deutsches Institut fur Normung (see Chapter 8) describes the business benefits of standards (DIN 2001). The study, conducted by the Technical University Dresden (TUD) and the Fraunhofer Institute for Systems and Innovation (SI), surveyed 4,000 firms in Germany, Austria, and Switzerland. The study confirmed that transaction costs are reduced when standards are used and that companies involved in standards work are more competitive and better equipped to meet market demands in new technologies. The study reveals the particular benefits to the firms of using International and European standards. The use of standards to make procurement more flexible (not confined to a single supplier) was emphasized. The use of standards to reduce product liability risk was also discussed (see Chapter 10). This study also pointed out that research risks are reduced by the use of standards by companies.

In a broad-based review of industry standards, American Society for Testing and Materials International (ASTMI) recently reported on both the qualitative and quantitative effects of standards on corporate profitability (Cendrowska 2007). The qualitative aspects were investigated by researchers at the University of Texas at Austin in their study of standardization in six sectors including small, medium, and large business organizations. This study was the basis for the study of the quantitative aspects of company standards performed by the firm Market Measurement. The qualitative and quantitative phases of the project showed the unmistakably positive aspects of company standards, including the return on investment of standards activities. One of the benefits listed was a reduced product liability exposure.

In the ICT area, Graham Bird (1998) describes the business benefits of *open standards*. Bird touts the flexibility, freedom of choice, reduced integration costs, and other benefits to both producers and users of such standards. Detailed analyses of standards in ICT appear in "Section VI: The Economic Perspectives" of *Advanced Topics in Information Technology Standards and Standardization Research* (Jacobs 2006) . These papers cover the subjects of network externalities and diffusion theory.

Variety Reduction (Simplification) Remains an Important Principle

It seems rather clear that there are still abundant opportunities for variety reduction. For example, many managers opposed the use of ISO 9000-based standards on quality management on the principle that "ours are better." They missed the point that the use of a common (variety reduced) quality management standard simplified the training of quality managers, engineers, and technicians, and also resulted in substantial progress in the *accreditation* of quality management systems.

As another example, consider the benefits of standard terminologies. The many conflicting definitions of a standard have greatly impeded communication and education in standards. A simple variety reduction of the term "standard" could rely on the ISO/IEC Guide 2 definition (see Chapter 1) with only small modifications *as needed* to adapt the term to special situations. The same principle could be applied to many other terms involved in standards and conformity assessment as outlined in Chapters 1 and 2.

The large variety of Web security standards makes the selection of suitable standards problematic for both industry and government. A reduction in the number and variety of standards developers, including consortia, would improve the coordination of those groups in avoiding duplication. As another mundane example, it has been reported that a Detroit automaker has a line of cars with eighty-one choices in side-view mirrors, while Honda has only two (Krisher 2006). Another example cited by Krisher states that another U.S. automaker uses one hundred varieties of catalytic converter, while an efficient competitor has only five.

Can variety reduction be carried too far? It appears that there are some "good varieties" and some "bad varieties" involved in this issue. Dominique Foray (1998) reports on that issue. An example is the maintenance of variety in computer security systems so that a single hack does not compromise the whole system. Another example is the selection of seed for crops. Some seeds of unused varieties of crops (corn, wheat, etc.) should be stored and retained so that they can be used if an unforeseen genetic-based malady hits the selected seed of choice. Foray also shows how standards can support innovation.

Relationship of Standards and Conformity Assessment to Innovative and Competitive Strategies

Innovation means bringing something to market that is new, whether or not it rises to the dignity of an invention. Invention, on the other hand, means the creation of something new, useful, and not obvious but does not necessarily involve bringing that creation to market. Some writers consider the whole process of innovation as including the invention. It is generally accepted that a standard can stifle or promote innovation. It is commonly thought that a standard that is born too early cuts off subsequent innovation. This effect can be minimized if the standard is revised frequently to accommodate innovations and is a performance-based standard rather than a "design standard." In some fields, especially in ICT, many innovations rest on an infrastructure of standards. As a result, the *lack* of a suitable standard may be costly and inhibit innovation. In ICT fields, an anticipatory standard may be required to field an innovation since compatibility and interoperation of such new developments is usually a requirement. A standard that arrives late may raise the "switching costs" of earlier innovations, that is, the extra costs of using the innovation after the fact may inhibit its use. It is often assumed that a standard that arrives "just-in-time" may avoid these effects. Innovation may also be inhibited by an inappropriate variety reduction. Such considerations show that policies on innovation must be flexible to meet rapid change. Ken Krechmer and Elaine Baskin (2007) have shown how innovation restricting ICT Regulations can be replaced by consensus adaptability standards that are innovation friendly. The impacts of standards and conformity assessment on innovation are explored in more detail in chapters 2 and 18 of the book *Standards, Innovation and Competitiveness* (Hawkins et al. 1995). Innovation, described somewhat differently than defined earlier, is the subject of a paper by Torsten Bahke (2007), the director of DIN. Bahke touches on the economics of standardization before focusing on the subject of standards and innovation. Bahke shows that standards are important at every phase of research and development in reducing uncertainty, and briefly discusses the relationship between standards and patents. Bahke then describes the current standards-related projects in Germany.

Competitiveness is usually considered to be at the heart of standards strategies for manufacturing and service firms. Hunter (1992) has shown how the use of specific National and International Standards enhances the competitiveness of a company. Hunter recommends the use of appropriate standards throughout the firm—in research and development, in marketing, in design and development, in purchasing, in manufacturing, and finally in installation and services. Companies like Ameritech, Ford, and AMP have established "strategic standards departments" in order to become more competitive in domestic and international trade (Howie 1995). Laura Hitchcock (2002) has described strategies for the whole aerospace industry. Pamela Suett (2004) of the American National Standards Institute (ANSI) describes a company-based virtual workshop to help firms develop and implement a strategic standardization plan for company members of ANSI. Michael G. Gorman (1995) outlines the strategic standards management concepts employed at Ameritech and Henry Line (1993) has reported on similar programs at AMP. Of course, the standard that would probably contribute the most to the competitiveness of U.S. industry is the International System of Units (SI), the modern metric system (Taylor 2001).

Strategic Standardization Management (SSM)™

Building on the concepts previously outlined, a *strategic standards management* discipline has evolved (Betancourt and Walsh 1995). The roots of the strategic standards management concept are apparently in an IEEE Computer Society Working Group (Camp 1990). Strategic Standardization Management (SSM)™ grew out of an ANSI program where the practices of companies like Ameritech, AMP, Deere & Company, Ford, Hewlett-Packard were analyzed to determine the best management practices in standardization at the company level (Walsh 1997). These concepts for standards management across the company were extended to apply to industries as well. The essential point of SSM is the coordination of the standardization activities throughout the company, including the application of suitable metrics for evaluating standards practices and determining if the return on investment (ROI) is adequate (Forselius 2003). The shift in viewpoint is to the *management* aspects of the standardization processes, including strategies for anticipatory standards and the globalization of competition. A major goal of SSM is to avoid the "ISO 9000 surprise" suffered by many U.S. firms. A company strategic standardization office was proposed to interface with senior management instead of the older practice of standards interfacing with other organizational entities. The concept for the strategic management of standards has been adopted by NIST (Leuteritz 1999).

Quality and Environmental Management Standards: ISO 9000 and ISO 14000

As indicated previously, the many good treatments of the use of quality and environmental management standards has been covered by many authors. Managers would do well to view the positive aspects of the use of these *management standards* in enhancing the performance of their companies. The use of these standards is especially important for firms engaged in international trade. As indicated earlier, many management standards are based on these International Standards. ISO 13485, for example, is a quality management system for the medical device industry and ISO 16949 is a quality management system for the auto industry.

Company Use of "External" Standards

In the past it has been common for firms to partition standards into "internal" and "external," especially when many of a firm's standards were internally developed. This distinction is now less useful since many National, Regional, and International Standards are used in the businesses of many companies. Mary Saunders (1997) at NIST has pointed out that the adoption of International Standards (simplification) and other measures can reduce the "transaction costs" of international trade for both large and small firms. The use of business-to-business (B2B) standards in electronic commerce (e-commerce) is surveyed by Aklouf et al. (2005), including the use of ISO standards (ISO 13584-25, 2004 and related ISO standards) for the PLIB or Parts Library standardization initiative. The Aklouf et al. paper includes a description of the electronic business XML (ebXML) by two large user organizations: UN/CEFACT (the United Nations Body for Trade Facilitation and Electronic Business) and the Organization for the Advancement of Structured Information Standards (OASIS). In addition, the ISO/IEC Joint Technical Committee 1 (see Chapter 6) has developed a series of standards ISO/IEC 15944 aimed at providing standards for e-business.

Wai Fong Boh et al. (2007) point out that B2B e-commerce systems must be designed to enhance the diffusion of the standards on a worldwide basis, taking into account the local variations needed to make such systems work.

Why a Company Should Participate in Standards Development

There are many good reasons for company participation in standards setting. The coordination benefits of such activity may be as valuable as the standard itself. The avoidance of a "standards surprise" may well be worth the effort, and, of course, the opportunity to shape the standard can be invaluable. John Hurd and Jim Isaak (2005) present good rationales for provider, user, government, and individual incentives for working in information technology standards development. The specific standards mentioned by Hurd and Isaak include the IEEE 802 (LAN) standards, the POSIX/UNIX standards, and the Linux standard. Individual attendance at meetings (real or virtual) is also a good technique for training in standards and conformity assessment (discussed in more detail in the following).

Managing Complexity in Conformity Assessment

As technology increases in complexity, the conformity assessment processes have also become more complex and costly. Specific examples were described in Chapter 7 on the interoperability testing of ICT equipment at the European Telecommunications Standards Institute (ETSI). Another example is the specialized Agilent test equipment needed to test WiMAX systems (Frenzel 2007). Consortia generally shift the burden of "operability" testing to the proposer of new standards (W3C and IETF). Metrology (measurement) standards have also become more complex to support modern systems such as the global positioning system. NIST plays a crucial role in these areas, establishing state-of-the-art measurement systems and transferring that technology to the private sector.

Macroeconomics of Standards

Part B of the DIN Summary Report (DIN 2001) referenced to earlier addresses the *macroeconomic* aspects of standards in Germany. Germany is dependent on foreign trade. The report emphasizes that innovation is a necessary, but not a sufficient, condition for the maintenance of competitiveness. Technology progress was estimated on the number of patents and level of research and development spending. The level of standardization was estimated via the number of published standards in the Perinorm© proprietary database and the International Classification for Standards (ICS) subject classification of standards (see Chapter 6). The study noted that new standards were published more frequently in innovative sectors. Standards were found to have a positive (supporting) effect on innovation. Since standards are vehicles for technology transfer, study respondents felt that technology leaders should become more active in standardization.

The national significance of standards, as compared with the contributions of capital, labor, and other factors, was determined by a regression analysis covering the period 1960–1966. The study found that standards were as significant to economic growth as patents. Foreign trade was found to be enhanced by standards by examining thirty-six bilateral trade relations for the year 1995. Standards were found to play a positive role in one-third of the technical sectors that had trade surpluses. The proposition that international standards lead to international competitiveness found empirical support in the study.

Part B of the study included a check on the consistency of its findings with the company standards part of the report (Part A). The result of the macroeconomic analyses showed that standards accounted for approximately 1 percent of the gross national product. The summary report concludes with a number of practical examples showing the positive effects of standards.

The UK Department of Trade and Industry (DTI) Economics Paper No. 12 (DTI 2005), with help from the British Standards Institution (BSI), empirically studied the economics of standards via three projects:

Project 1: Standards and Long-Run Growth in the UK

Project 2: The Impact of Standards on Productivity in Manufacturing

Project 3: Do Standards Enable or Constrain Innovation?

This study made use of some of the concepts (use of the Perinorm and BSI Catalogs) used in the earlier DIN study.

One of the results of Project 1 was that standards accounted for approximately 13 percent of the growth in labor productivity in the UK over the period 1948–2002 as measured by the growth of standards in the BSI Catalog and a calculation of the elasticity of labor productivity for each 1 percent growth of standards. Since many consortia and de facto standards do not make it into the BSI Catalog, the 13 percent figure may be too low (author's comment). The study estimated that standards contribute approximately $4.4 billion to the UK economy (at then-existing exchange rates) and that the return from investments in standards made good business sense at both macroeconomic and microeconomic levels.

The Project 1 study referred to some results of Gregory Tassey (1995), which estimated the economic benefit (called "social rate of return" or SSR by economists) to the U.S. industry of the measurement standards produced by NIST.

Project 2 expanded the scope of the study in Project 1 to include three other European countries: France, Germany, and Italy. Germany was found to have the largest stock of standards. The ongoing "dilution" of the stock of BSI standards due to adoption of European standards to replace national standards (see Chapter 7) was studied for its "technology transfer" implications. Otherwise, the results of Project 2 confirmed the results of Project 1.

Part 3 focused on the work of professor G. M. P. Swann. It was designed to empirically test some of the earlier theoretical work of Swann. The result was that standards of medium vintage usually enhance innovation, whereas older and newer standards may actually inhibit innovation.

A review of the literature of economics and standards has revealed only one study of the United States or North America like the DIN or DTI studies. That study, performed by Joseph Haimowitz and Joanne Warren (2007) of the Conformance Board of Canada under contract to the Standards Council of Canada, builds on the European studies described earlier. This report focuses on the economic effects of standards on Canada's economy and includes a literature review including the effects of standards on international trade and economic growth as well as the effects on productivity. The study included summaries of interviews with prominent Canadian industry people and also included two case studies of standards used by Canadian industrial firms.

Access to Standards and Standards Developers

It is a historical fact that until recently, access to standards was a major difficulty for those faced with the identification and acquisition of standards (Spivak and Winsell 1991). The rapid development of ICT and the modernization of SDO standards distribution have now made those problems *manageable.* ISO and UNESCO (1986) produced a document that provided guidance on the subject of access that still contains some useful information on these subjects. Many authors, including Clifford Lynch (1999) have criticized the economic model of SDOs that require clients to purchase copyrighted copies at relatively high prices. It is not productive to rehash the arguments here, but a paper by Rada and Berg (1995) sums up the issues. Of course, many consortia standards are published free on the Web.

If you work for a substantial company, a good place to start accessing standards is the "Corporate Librarian." Nancy C. Taylor (1991) points out that librarians are the liaison between standards producers and standards users. Librarians and the members of the Special Libraries Association (SLA, based in Washington DC) have recently focused on standards. Some special librarians have become members of the Standards Engineering Society to improve their knowledge of standards and share their expertise with the rest of us.

Search Resources

A good place to search for standards is ANSI's National Standards Systems Network (NSSN), the "search engine for standards." One can search using a title, an abstract, a keyword or a Standard number. The NSSN is at the Web address www.nssn.org. ANSI supplies some background on the NSSN and some tips for enhancing searches. Paul W. Mercer (1995) describes the historical development, including the funding by the Advanced Research Projects Agency (ARPA), who brought us the TC/IP invention.

The NSSN Web site includes links to Government Agency Web sites, a list of U.S. standards developers, and an Email Standards Tracking and Automated Reporting (STAR) service. ANSI also provides a Site License for downloadable standards that makes them available 24/7. ANSI also publishes the weekly *Standards Action* online that is valuable for keeping up with new standards.

Another good resource for finding standards information is the National Center for Standards and Certification Information (NCSCI) at the National Institute of Standards and Technology (NIST) in Gaithersburg, Maryland. NCSCI is the WTO contact point required by the Agreement on Technical Barriers to Trade (TBT) and is also the U.S. Member of the ISO Information Network (ISONET). The NCSCI collection includes copies of the *Official Journal of the European Communities* where EU Directives are published. JoAnne Overman (1997) describes the NCSCI services in more detail. The URL for the NCSCI is ts.nist.gov/ncsci.

Many Sources, Many Options for Delivery

Most standards developers have a convenient inquiry Web site. The cost of membership in many standards organizations (e.g., ASTMI, IEEE, SES, and ANSI) is very reasonable. The cost of some standards is thus reduced. One of the convenient ways to keep up to date on standards development is to join a trade association. Some SDOs (ASTMI and IEEE) will publish "special collections" of related standards at reduced cost over individual copies. Standards can be obtained on virtually all the modern media (online, CD-ROM, etc.) and ASTMI has just announced that its standards will soon be available in HTML, facilitating searching via links. ASTMI online standards will continue to be available in PDF format as well. Scopes of ASTMI standards can be searched at www.astm.org/dsearch.htm. The availability of digital libraries for both standards and other technical literature is growing (www.astmdigitallibrary.org). A useful tool for researchers and librarians is the ISO Key-Word-In-Context (KWIC) Index of International Standards. DIN has opened an "Info-Point" in the United States because the use of DIN standards in the United States are the third most commonly used (after ASTM and ISO standards). The Info-Point may be reached by contacting din@uschamber.com.

Access to SDOs and Consortia

The expenses of participation in SDOs and Consortia vary. There are frequently "tiers" of participation where higher levels of influence cost more, especially in consortia. Membership in an SDO is frequently a requirement and the principle is "pay to play." Access to national and international standards bodies is pretty routine, but access to some regional standards bodies (e.g., the EU standards bodies) is more problematic unless you are a multinational corporation with a presence there.

U.S. access to international standards and conformity assessment activities has been an ongoing concern. Every few years there is a big meeting to discuss the issues. Two major publications summarizing these issues have been released by ANSI (ANSI 1990, 1996). One modest suggestion to improve the situation would be that we start using the language of International Standards. The IEC has granted special access to the IEEE for the adoption of its standards and ETSI has opened its door to U.S. participants so as to not lose out on advanced U.S. technical developments.

Management Aspects of Accreditation

To say that the management aspects of accreditation involve high ethical and other standards related to accreditation is to state the obvious. The careful conformance to standards such as ISO/IEC 17011 (outlined in Chapter 5) is a necessity, especially for accreditations involving international conformity assessments. The roles of the International Accreditation Forum (IAF) and the International Laboratory Accreditation Cooperation (ILAC), covered in Chapter 6, should be understood by management. The reach of an accreditor, through Mutual Recognition Agreements (MRAs) and other mechanisms, is an important aspect of selecting an accredited Conformity Assessment Body (CAB). The ANSI-ASQ (ANAB) covered in Chapter 8 and the federal accreditations (NVLAP and NVCASE) covered in Chapter 11 are vitally important aspects of this subject. Laboratory accreditation is expensive but important. The Houston, Texas, (Police) Crime Lab, which made life and death assessments, had no accreditation! The same is, unfortunately, true of many other laboratories and CABs.

Standards Education: A Management Imperative

It's the two cultures again! Managers need to make sure they are educated in standards and provide for the education of employees in standards. Education is another of the long-running issues in standards; it formed a significant part of a source book edited by Spivak and Winsell (1991). Most of us in standards work received our education the hard way (by experience). The dearth of college level-courses in standards emphasized in Spivak and Winsell was found to be a continuing problem. Focusing on standards education for engineers, William E. Kelly (2001), dean of engineering at Catholic University of America (CUA), pointed out ten years later that standards education was an accreditation requirement for schools of engineering. Donald Purcell at CUA reported in 2003 that only a handful of universities actually offered such courses. Purcell summarized the comments of the participants in the CUA survey in the May 2003 issue of *Standardization News*. That elicited comment from Steven Spivak who charged the SDOs, ANSI, and other professional groups to help fix the problem. A number of papers on the use of standards in university engineering courses have been published (Abran 1996; Jenkins 1999). Abran describes the teaching of software engineering by using ISO standards. Jenkins outlines a hands-on approach to teach standards and codes in mechanical engineering.

Many educators, like the author, have used standards in classroom work. It appears that a good plan for inserting standards issues in university education would be to have various appropriate standards included in classroom work throughout undergraduate education followed by the "capstone course" in design that involves standards. The high cost of standards has been an impediment to the use of standards in engineering education. That issue has been alleviated somewhat by the use of a site license, like that at the University of Texas at Austin in the Engineering Library. ASTMI has also arranged to provide low-cost copies of standards for classroom work (Schultz 2005).

ASTMI held a roundtable on standards education in the United States in June 2003, at which nine leaders of the standards community discussed the issues. As more attention was focused on the issue, it started to get more traction in ANSI, ASTMI, IEEE, and other venues. Pamela Suett (2004) of ANSI described ANSI's new online education programs in standards education. These courses supplement classroom education in standards that ANSI had historically sponsored, some in cooperation with SES. ANSI's e-learning programs are available at www.StandardsLearn.org. Other organizations, notably IEEE, got into the act with standards education programs (Bickart 2007). Bickart's paper emphasized the fact that the National Standards Strategy for the United States (2000) and the United States Standards Strategy published in 2005 emphasized the need for standards education. The IEEE Standards in Education Portal is described in Bickart's paper, which announced the award of a three-year National Science Foundation (NSF) grant to establish a pilot test program. Standards education initiatives were announced by eight prominent SDOs. It looks like the tide may be turning on standards education; a special issue of the *International Journal of IT Standards and Standardization Research* was recently published (Vol. 5, No. 2, July–December 2007).

Evidently, the management/business schools need to get with the standards education program, too. A 2001 text on "Global Business" barely mentions standards.

References

Abran, Alain, 1996, Teaching Software Engineering Using ISO Standards, *ACM StandardView* 4(3): 139–145.

Aklouf, Youcef, Guy Pierra, Yamine Ait Ameur, and Habiba Drias, 2005, PLIB Ontology: A Mature Solution for Products Characterization in B2B Electronic Commerce, *International Journal of IT Standards and Standardization Research* 3(2): 66–81.

American National Standards Institute (ANSI), 1990, ANSI Analysis of National Institute of Standards and Technology Office of Standards Services Hearing on Improving U.S. Participation in International Standards Activities, New York, American National Standards Institute.

American National Standards Institute (ANSI), 1996, American Access to the European Standardization Process, New York, American National Standards Institute.

Bahke, Torsten, 2007, Innovation and Standardization, *SES Standards Engineering* 59(6): 1–7.

Betancourt, Diego, and Robert Walsh, 1995, The Evolution of Strategic Standardization Management (SSM), *ACM StandardView* 3(3): 117–126.

Bickart, Theodore A., 2007, An IEEE Perspective on Standards in Education, *SES Standards Engineering* 59(1): 1–9.

Bird, Graham B., 1998, The Business Benefit of Standards, *ACM StandardView* 6(2): 76–80.

Boh, Wai Fong, Christina Soh, and Steven Yoh, 2007, A Case Study of RosettaNet, *Communications of the ACM* 50(12): 57–62.

Camp, Clyde R., 1990, Why TI (Texas Instruments) Needs a Standards Strategy, *TI Technical Journal* 7(3): 2–15. (Note: A modified version of this paper appeared in the December 1997 issue of *ASTM Standardization News*.)

Cendrowska, Teresa J., 2007, The Impact of Standards on Corporate Profitability, *ASTM Standardization News* 35(7): 19-21.

Department of Trade and Industry (DTI), 2005, *DTI Economics Paper No. 12: The Empirical Economics of Standards*, London, Department of Trade and Industry (DTI) and British Standards Institution (BSI).

Deutsches Institut fur Normung (DIN), 2001, *Economic Benefits of Standardization, Summary of Results, Final Report and Practical Examples, Part A: Benefits for Business*, Berlin, DIN German Institute for Standardization e.V., pp. 1–19.

Ferrell, Joseph, 1990, The Economics of Standardization: A Guide for Non-Economics, in *The Information Technology Standardization Process*, edited by John L. Berg, Amsterdam, Elsevier Science/North-Holland, pp. 189–198.

Foray, Dominique, 1998, Standards and Innovation in Technological Dynamics, *ACM StandardView* 6(2): 81–84.

Forselius, Richard J., 2003, The Need for Effective Standardization Management, *ASTM Standardization News* 31(4): 32–33.

Frank, Robert H., and Ben S. Bernanke, 2004, *Principles of Microeconomics*, 2nd ed., New York, McGraw-Hill/Irwin, p. 15.

Frenzel, Louis E., 2007, WiMAX: The Real Deal, *Electronic Design* 55(24): 46.

Gallagher, Scott, 2007, The Complementary Role of Dominant Designs and Industry Standards, *IEEE Transactions on Engineering Management* 54(2): 371–379.

Gorman, Michael G., 1995, Strategic Standards Management at Ameritech, *SES Standards Engineering* 47(3): 1–5.

Haimowitz, Joseph, and Joanne Warren, 2007, *Economic Value of Standardization*, submitted to the Standards Council of Canada by The Conference Board of Canada, www.scc.ca.

Hawkins, Richard, Robin Mansell, and Jim Skea, eds., 1995, *Standards, Innovation and Competitiveness: The Politics and Economics of Standards in Natural and Technical Environments*, Brookfield, VT, Edward Elgar.

Hemenway, David, 1975, *Industrywide Voluntary Product Standards*, Cambridge, MA, Ballinger.

Hitchcock, Laura E., 2002, Standards During Times of Change: Aerospace Strategies for Keeping Standards and Business Linked, *SES Standards Engineering* 54(6): 1–6.

Howie, Robert L. Jr., 1995, Competing through Standardization, Special Advertising Section of *Business Week*, October 16.

Hunter, Robert, 1992, Standards Enhance Competitiveness, *SES Standards Engineering* 44(6): 133–137.

Hurd, John, and Jim Isaak, 2005, IT Standardization: The Billion Dollar Strategy, *International Journal of IT Standardization and Research* 3(1): 68–74.

ISO/UNESCO, 1986, *Access to Standards Information: How to Inquire or Be Informed about Standards and Technical Regulations Available Worldwide*, Geneva, International Organization for Standardization.

Jacobs, Kai, ed., 2006, *Advanced Topics in Information Technology Standards and Standardization Research*, vol. 1, Hershey, PA, Idea Group, pp. 251–319.

Jenkins, Michael G., 1999, Standards and Codes in Mechanical Engineering Education, *ASTM Standardization News* 27(9): 20–25.

Kelly, William E., 2001, Including Standards in the Education of Future Engineers, *SES Standards Engineering* 53(1): 1–5.

Krechmer, Ken, and Elaine Baskin, 2007, The Entrepreneur and Standards, *SES Standards Engineering* 59(2): 1–8

Krisher, Tom, 2006, Report: Inefficiency Plagues U.S. Carmakers, *Austin American-Statesman*, October 3, p. C3.

Leuteritz, Krista Johnsen, 1999, Towards Strategic Management of Standards Activities at NIST, *ASTM Standardization News* 27(12): 26–29.

Line, Henry, 1993, Industry Standards: A Key Factor to Market Success, How AMP Addresses This Critical Element, *ASTM Standardization News* 21(11): 32–41.

Lynch, Clifford, 1999, The Case for New Economic Models to Support Standardization Efforts, *SES Standards Engineering* 51(1): 1–6.

Mercer, Paul W., 1995, The National Standards Systems Network (NSSN), *ASTM Standardization News* 23(12): 26–29.

Miller, James C., 1990, Rating the Effectiveness of Your Standards Program, in *Standards Management: A Handbook for Profits*, New York, American National Standards Institute, 421–427.

Overman, JoAnne, 1997, NIST's Information Center ... Helping Users Find Standards, *SES Standards Engineering* 49(2): 14–15.

Purcell, Donald E., 2003, Standards Education Survey, *SES Standards Engineering* 55(4): 17–18.

Rada, Roy, and John Berg, 1995, Standards: Free or Sold? *Communications of the ACM* 38(2): 23–27.

Saunders, Mary, 1997, Reducing the Transaction Costs of International Trade, *ASTM Standardization News* 25(6): 17–19.

Schultz, Daniel, 2005, Standards in the Classroom, *ASTM Standardization News* 33(7): 20–23.

Spivak, Stephen M., and Keith A. Winsell, eds., 1991, *A Sourcebook of Standards Information: Education, Access and Development*, Boston, G.K. Hall & Co.

Suett, Pamela, 2004, The Changing Options for Delivering Standards Education, *SES Standards Engineering* 56(2): 1–7.

Tassey, Gregory, 1995, The roles of standards as technology infrastructure, in *Standards, Innovation and Competitiveness*, edited by Richard Hawkins, Robin Mansell, and Jim Skea, *The Politics and Economics of Standards in Natural and Technical Environments*, Brookfield, VT, Edward Elgar, pp. 161–171.

Taylor, Barry N., ed., 2001, *The International System of Units (SI)*, NIST Special Publication 330, Gaithersburg, MD, National Institutes of Standards and Technology.

Taylor, Nancy C., 1991, The Corporate Librarian: Liaison Between Standards Producers and Users, *SES Standards Rngineering* 43(1): 10–12.

Toth, Robert, ed. 1984, *The Economics of Standardization*, Miami, FL, The Standards Engineering Society.

Toth, Robert, ed., 1990, *Standards Management: A Handbook for Profits*, New York, American National Standards Institute.

Walsh, Robert 1997, Strategic Standardization Management Emerges as a Key Business Strategy, *ANSI Reporter*, pp. 9–12.

Appendix 1: Alphabetical Index for Chapter 1 (Guide 2)

As pointed out in Chapter 1, the terms and definitions of ISO/IEC Guide 2 are groups of related concepts under generic clause headings. This arrangement is most useful in showing the relationships of the terms and definitions. The alphabetical index of Guide 2, shown in this Appendix 1, provides a convenient way of finding a term and its definition after the structure has been forgotten. The numbers refer to the clause numbers in Guide 2. The entries are boldface to correlate with the boldface entries in Chapter 1.

Although only the English (E) terms are presented here, the three-language index in Guide 2 provides some "translated terminology" that is useful. Additional terms in five other languages, sans definitions, are also useful in that regard.

Appendix 2: Alphabetical Index for Chapter 2 (ISO/IEC 17000)

As was the case in Chapter 1, the terms and definitions in ISO/IEC 17000 are grouped by related concepts under generic clause headings. That arrangement is useful in showing the relationships of the terms and definitions. The alphabetical index of IS 17000, shown in this Appendix 2, provides a convenient way of finding a term after the structure has been forgotten. As before, the numbers refer to the clause numbers in IS 17000 and the entries are boldface to correlate with the showing in IS 17000. The terms and their definitions appear in ISO/IEC 17000 in English and French, and the Informative Annex A appears in English, French, and Russian. As before, only the English terms and definitions appear in Chapter 2.

Appendix 3: Abbreviations and Acronyms for Standards, Conformity Assessment, and Accreditation

Note: Additional abbreviations appear in http://ts.nist.gov/ts/htdocs/210/ 216/acro1.htm. Many of the abbreviations and acronyms listed here are not discussed in the text but frequently appear in the references cited and in other literature. More information on all of these items can be obtained from the usual search engines.

A2LA	American Association for Laboratory Accreditation
AAMI	Association for Advancement of Medical Instrumentation
AASHTO	American Association of State Highway and Transportation Officials
ACCSQ	ASEAN Consultative Committee on Standards and Quality
ACIL	(formerly) The American Council of Independent Laboratories
ACM	Association for Computing Machinery
ACTA	Administrative Council for Terminal Attachments
AEA	American Electronics Association
AG	CEN General Assembly
AIA	Aerospace Industries Association of America
AIC	ANSI International Committee
AIF	ANSI International Forum, part of the ANSI International Committee
ANAB	American National Accreditation Board
ANS	American National Standards
ANSI	American National Standards Institute
APA	Administrative Procedures Act
APEC	Asia Pacific Economic Cooperation
API	American Petroleum Institute
APLAC	Asia Pacific Laboratory Accreditation Cooperation
APLMF	Asia-Pacific Legal Metrology Forum
APMP	Asia Pacific Metrology Program
APT	Asia Pacific Telecommunity
ASC	Accredited Standards Committee

ASD	Accredited Standards Developer
ASEAN	Association of Southeast Asian Nations
ASME	American Society of Mechanical Engineers
ASQ	American Society for Quality
ASTMI	American Society for Testing and Materials International
AT&T	American Telephone and Telegraph
ATIS	Alliance for Telecommunications Industry Solutions
ATM Forum	Asynchronous Transfer Mode (Consortium)
ATSC	Advanced Television Systems Committee
ATSC	Australian Telecom Standards Committee
BDT	ITU Telecommunications Development Bureau
BIPM	International Bureau of Weights and Measures
BOCA	Building Officials and Code Administrators
BR	Radio Communications Bureau (ITU-R)
BOCAI	Building Officials and Code Administrators International, Inc.
BPVC	Boiler and Pressure Vessel Code
BR	Radio Communication Bureau
BSI	British Standards Institution
BSR	Board of Standards Review (ANSI)
BT	CEN Technical Management Board
CA	Conformity Assessment
CA	CEN Administrative Board
CAB	Conformity Assessment Body
CAB	Conformity Assessment Board (IEC)
CAC	Codex Alimentarius Commission
CAG	Chairman's Advisory Group (CASCO)
CANENA	Council for Harmonization of Electrotechnical Standards of the Nations of the Americas
CAPC	Conformity Assessment Policy Committee (ANSI)
CASCO	ISO Committee on Conformity Assessment
CB FCS	CB Full Certification Scheme (extension of CB Scheme)
CB Scheme	IECEE Scheme for Recognition of the Results of Testing Electrical Equipment
CCAF	CENELEC Conformity Assessment Forum
CCIR	International Radio Consultative Committee
CD	Committee Draft
CDMA	Code Division Multiple Access

CEA	Consumer Electronics Association
CE Mark	EU mark indicating conformance to Essential Requirements of an EU Directive
CEC	Commission of the European Community
CEN	European Committee for Standardization
CENELEC	European Committee for Electrotechnical Standardization
CEC	Canadian Electrical Code
CFR	Code of Federal Regulations
CGSB	Canadian General Standards Board
CIGRE	International Conference of Large Electric Systems
CISPR	International Special Committee on Radio Interference
CITEL	Inter-American Telecommunications Conference
CMC	CEN Management Center
CMC	Certification Management Committee (IEC, Manages CB Scheme)
CMC	Company Member Council (ANSI)
COPANT	Pan American Standards Commission
COPOLCO	ISO Committee on Consumer Policy
COS	Corporation for Open Systems International
CPSA	Consumer Product Safety Act
CPSC	Consumer Product Safety Commission
CSA	Canadian Standards Association
CTI	Comparative Tracking Index
CTIA	Cellular Telecommunications Industry Association
CWA	CEN or CENELEC Workshop Agreement
DG	Directorate General (EU)
DEVCO	ISO Committee on Developing Country Matters
DIN	Deutsches Institut fur Normung
DIS	Draft International Standard
DOC	Declaration of Conformity
DOC	Department of Commerce (U.S.)
DOD	Department of Defense (U.S.)
DOE	Department of Energy (U.S.)
DOJ	Department of Justice (U.S.)
DOL	Department of Labor (U.S.)
DOT	Department of Transportation (U.S.)
DSDV	Digital Simultaneous Voice and Data Forum

DTI	Department of Trade and Industry (UK)
EA	European Cooperation for Accreditation
EBU	European Broadcasting Union
EC	European Community, now the European Union (EU)
ECCB	Electronic Components Certification Board
ECMA	(formerly) European Computer Manufacturers Association
ECSA	Exchange Carriers Standards Association
ECSC	European Coal and Steel Community
EEA	European Economic Area
EEC	European Economic Community
EFTA	European Free Trade Association
EIA	Electronic Industries Association
EMC	Electromagnetic Compatibility
EN	European Norm
ENV	European Pre-Standard
EOTA	European Organization for Technical Approvals
ETSI	European Telecommunications Standards Institute
EU	European Union, previously the European Community (EC)
EWOS	European Workshop for Open Systems
FCC	Federal Communications Commission
FDIS	Final Draft International Standard
FIPS	Federal Information Processing Standards
FQA	Fastener Quality Act
FQA99	Fastener Quality Act Amendment Act
FR	Federal Register
FSAN	Full Service Access Network
FTAA	Free Trade Area of the Americas
FTC	Federal Trade Commission
GATT	General Agreement on Tariffs and Trade
GII	Global Information Infrastructure
GMC	Government Member Council (ANSI)
HACCP	Hazard Analysis Critical Control Point
HD	Harmonized Document (two definitions in Chapter 3 and Guide 2)
HDTV	High-Definition Television
IAAC	Inter-American Accreditation Cooperation
IAF	International Accreditation Forum

IAN	Industry Advisory Network (ISO-CASCO)
IBC	International Building Code
ICAC	International Conformity Assessment Committee (ANSI)
ICAO	International Civil Aviation Organization
ICBO	International Conference of Building Officials
IC	International Committee (ANSI)
ICC	International Code Council
ICCEC	ICC Electrical Code Administrative Provisions
ICRU	International Commission on Radiation Units and Measurements
ICS	International Classification for Standards
IEC	International Electrotechnical Committee
IECC	International Energy Conservation Code
IECEE	IEC System for Recognition of Results of Testing to Standards for Electrical Equipment (see CB Scheme)
IECEx	IEC Scheme for Electrical Equipment Intended for Use in Explosive Atmospheres
IECQ	IEC Quality Assessment System for Electronic Components
IEEE	Institute of Electrical and Electronics Engineers
IEEE-ISTO	IEEE Industry Standards and Technology Organization
IETF	Internet Engineering Task Force
IEV	International Electrotechnical Vocabulary
IFAN	International Federation of Standards Users
IFC	International Fire Code
IFGC	International Fuel Gas Code
IFI	Industrial Fasteners Institute
IFRB	International Frequency Registration Board
ILAC	International Laboratory Accreditation Cooperation
IMC	International Mechanical Code
INCITS	InterNational Committee for Information Technology Standards
IPC	International Plumbing Code
IPMC	International Property Maintenance Code
IPR	Intellectual Property Right
IPSDC	International Private Sewage Disposal Code
IRC	International Residential Code
ISA	International Society for Measurement and Control
ISEAL	Social and Environmental Issues
ISO	International Organization for Standardization

ISOC	Internet Society
ISONET	ISO Information Network
ISO-SR	ISO Committee on Social Responsibility
ISSS	CEN Information Society Standardization System
ITA	Industry Technical Agreement, ISO/IEC normative document developed in a workshop outside the ISO/IEC structure but administered by an ISO/IEC designated body
ITIC	Information Technology Industry Council
ITL	Information Technology Laboratory (in NIST)
ITSC	Interregional Telecom Standards Conference
ITU	International Telecommunications Union (United Nations)
ITU-D	ITU Telecommunication Development Sector
ITU-R	ITU Radio Communication Sector
IZC	International Zoning Code
JEDEC	Joint Electronic Device Engineering Council (in EIA)
JISC	Japanese Industrial Standards Committee
JTC 1	ISO/IEC Joint Committee on Information Technology
KWIC	Key Word In Context
LAP	Laboratory Accreditation Program (NIST)
LVD	Low Voltage Directive
MIT	Massachusetts Institute of Technology
METI	Ministry of Economy, Trade and Industry (Japan, replaced MITI)
MIL SPEC	Military Specification
MIL STD	Military Standard
MITI	Ministry of International Trade and Industry (Japan)
MLA	Multilateral Agreement
MoU	Memorandum of Understanding
MPEG	Moving Pictures Expert Group
MRA	Mutual Recognition Agreement
MTBE	Methyl Tertiary Butyl Ether
MUPLA	Model Uniform Products Liability Act
NACLA	National Cooperation for Laboratory Accreditation
NAFTA	North America Free Trade Agreement
NAI	National Authorized Institution
NB	Notified Body

NCB	National Certification Body
NCSCI	National Center for Standards and Certification Information (in NIST)
NCSLI	National Conference of Standards Laboratories, International
NEC	National Electrical Code
NECQ	National Electronic Component Quality Assessment System
NEMA	National Electrical Manufacturers Association
NFDA	National Fastener Distributors Association
NFPA	National Fire Protection Association
NHTSA	National Highway Traffic Safety Administration (in DOT)
NISO	National Information Standards Organization
NIST	National Institute of Standards and Technology (in DOC)
NORAMET	North American Cooperation in Metrology
NP	New Work Item Proposal
NPC	National Policy Committee (ANSI, on Education)
NPRM	Notice of Proposed Rulemaking
NRC	National Research Council
NRTL	Nationally Recognized Testing Laboratory (OSHA)
NSB	National Standards Body
NSPI	National Spa and Pool Institute
NSSN	National Standards System Network (in ANSI)
NTIA	National Telecommunication and Information Administration (DOC)
NTIS	National Technical Information Service (DOC)
NTTAA	National Technology Transfer and Advancement Act
NVCASE	National Voluntary Conformity Assessment Systems Evaluation Program (in NIST)
NVLAP	National Voluntary Laboratory Accreditation System (in NIST)
NW	New Work
OAS	Organization of American States
OECD	Organization for Economic Cooperation and Development (UN)
OG	Open Group (Consortium)
OIF	Optical Internetworking Forum
OIML	International Organization of Legal Metrology
OMB	Office of Management and Budget
OSHA	Occupational Safety and Health Administration (DOL)

OSS	Office of Standards Services (NIST)
OTA	Office of Technology Assessment
OWM	Office of Weights and Measures (NIST)
PAC	Pacific Accreditation Cooperation
PAS	Publicly Available *Standard*, ISO normative document representing the consensus of an ISO working group (alternatively, depending on context, a Publicly Available *Specification*, produced by a consortium).
PASC	Pacific Area Standards Congress
PINS	Project Initiation Notification System
P.L.	Public Law
PSB	CEN Partner Standardization Bodies
PTO	U.S. Patent and Trademark Office (DOC)
PWI	Preliminary Work Item
QMS	Quality Management System (e.g., ISO 9000 family)
QPL	DOD Qualified Products Listing Program
QS	Quality System
QSR	Quality System Registrar
R&TTE	Radio & Telecommunications Terminal Equipment
RAND	Reasonable and Non-Discriminatory (Rates in Patent Licensing)
REMCO	ISO Committee on Reference Materials
RFC	Request For Comments
RFI	Radio Frequency Interference
RoHS	EU Restrictions on Hazardous Substances
RPM	Regional Preparatory Meetings (ITU-D)
RRB	Radio Regulations Board (ITU-R)
SAE	Society of Automotive Engineers
SBCCI	Southern Building Code Congress International
SC	Subcommittee (in a Standards Organization)
SCC	Standards Council of Canada
SCTE	Society of Cable Telecommunications Engineers
SDO	Standards Developing Organization
SDOC	Suppliers' Declaration of Conformity (First-Party Attestation)
SEMI	Semiconductor Equipment and Materials Institute, International
SES	Standards Engineering Society
SI	Supervising Inspectorates
SLA	Special Libraries Association

SRM	Standard Reference Material
SSC	Standard Setting Consortium
SSM	Strategic Standards Management
STAR	Standards Tracking and Automated Reporting (service)
TAG	Technical Advisory Group
TAP	Traditional Approval Procedures (ITU)
TBT	Technical Barrier to Trade
TC	Technical Committee
TIA	Telecommunications Industry Association
TREAT	Trans-Regional EU-ASIAN Trade Initiative
TSACC	Canada, Telecom Standards Advisory Committee
TSB	Telecommunication Standardization Bureau (in ITU-T)
TMB	ISO Technical Management Board
TTA	Korea, Telecom Technical Association
TTC	Japan, Telecom Technology Committee
UCC	Uniform Commercial Code
UL	Underwriters Laboratories
U.S.C.	United States Code
USNC	U.S. National Committee (of the IEC)
USSS	United States Standards Strategy
USTR	U.S. Trade Representative
VA	Vienna Agreement
VDE	Verband Deutscher Elektrotechniker (German Association of Electrical Engineers)
VIM	International Vocabulary of Basic and General Terms in Metrology
VPS	Voluntary Product Standard (NIST)
WARC	World Administrative Radio Conference
WD	Working Draft
WEEE	Waste Electrical and Electronic Equipment (EU)
WG	Working Group
WRC	World Radiocommunication Conference
WTO	World Trade Organization
WTDC	World Technical Development Conference (ITU-D)
WWW	World Wide Web

Appendix 4: URLs for Standards, Conformity Assessment, and Accreditation

A2LA (American Association for Laboratory Accreditation): www.a2la.org

ACTA (Administrative Council for Terminal Attachments): www.part68.org

AIA (Aerospace Industries Association): www.aia-aerospace.org

ANSI (American National Standards Institute): web.ansi.org

ANSI e-learning: www.StandardsLearn.org

APLAC (Asia Pacific Laboratory Accreditation Cooperation): www.aplac.org

APT (Asia Pacific Telecommunity): www.aptsec.org

ASEAN (Association of Southeast Asian Nations): www.aseanec.org

ASME (American Society of Mechanical Engineers): www.asme.org

ASTMI (American Society for Testing and Materials, International): www.astm.org

ATIS (The Alliance for Telecommunications Industry Solutions): www.atis.org

CEA (Consumer Electronics Association): www.ce.org

CEN (European Committee for Standardization): www.cenorm.be

CENELEC (European Committee for Electrotechnical Standardization): www.cenelec.be

CITEL (Inter-American Telecommunications Commission): www.citel.oas.org

Consortia (Andrew Updegrove): www.consortiuminfo.org

CSA (Canadian Standards Association): www.csa.ca

DIN (Deutsches Institut fur Normung): www.din.de

DIN (Deutsches Institut fur Normung, in the U.S.): din@uschamber.com (e-mail)

EA (European Co-operation for Accreditation): www.european-accreditation.org

ETSI (European Telecommunications Standards Institute): www.etsi.org

FIPS (Federal Information Processing Standard): csrc.nist.gov/piv-project/index.html

IAF (International Accreditation Forum): www.iaf.nu

ICC (International Code Council): www.iccsafe.org

IEC (International Electrotechnical Commission): www.iec.ch

IECEE (IEC System for Recognition of Results of Testing to Standards for Electrical Equipment): www.iec.ch

IECEE–CB scheme www.cbscheme.org

IECEx (IEC Scheme for Electrical Equipment Intended for Use in Explosive Atmospheres): www.iecex.com

IEEE (Institute of Electrical and Electronics Engineers): www.ieee.org

IETF (Internet Engineering Task Force): www.ietf.org

IFAN Online (International Federation of Standards Users): www.ifan-online.org

ILAC (International Laboratory Accreditation Forum): www.ilac.org

INCITS (InterNational Committee for Information Technology Standards): www.incits.org

IQNet (International Certification Network): www.iqnet-certification.com

ISEAL (social and environmental issues): www.isealalliance.org

ISO (International Organization for Standardization): www.iso.ch/

ISO/IEC www.standardsinfo.net

ISO-SR (ISO Committee on Social Responsibility): www.iso.org/sr

ISOC (Internet Society): www.isoc.org

ISSS (Information Society Standardization System): www.cenorm.be/isss

ITU (International Telecommunications Union): www.itu.int/aboutitu/index.html

JTC 1 (ISO/IEC Joint Technical Committee 1): www.jtc1.org

NACLA (National Cooperation for Laboratory Accreditation): www.nacla.net

NCSLI (National Conference of Standards Laboratories International): www.ncsli.org

NEMA (National Electrical Manufacturers Association): www.nema.org

News Standards Summit   www.newssummit.org

NIST   (National Institute of Standards and Technology): www.nist.gov

NIST   (checklists): csrc.nist.gov/checklists

OG   (Open Group): www.opengroup.org

PAC   (Pacific Accreditation Cooperation): www.apec-pac.org

PASC   (Pacific Area Standards Congress): www.pascnet.org

SCC   (Standards Council of Canada): www.scc.ca/en

SEMI   (Semiconductor Equipment and Materials International): www.semi.org

SES   (Standards Engineering Society): www.ses-standards.org

TIA   (Telecommunications Industry Association): www.tiaonline.org
(Underwriters Laboratories): www.ul.com

USSS   (United States Standards Strategy): www.us-standards-strategy.org

WWW   (World Wide Web): www.w3.org

Appendix 5: List of NIST Guides on Selected European Union Directives

Co-edited by Helen Delaney and Rene van de Zande

Downloadable from: http://ts.nist.gov/Standards/Global/cainfo.cfm.

NIST GCR 01-812	A Guide to the EU Low Voltage Directive
NIST GCR 01-813	A Guide to the EU Directive on Electromagnetic Compatibility
NIST GCR 01-814	A Guide to the EU Machinery Directive
NIST GCR 01-815	A Guide to the EU Medical Devices Directive
NIST GCR 01-816	A Guide to the EU Active Implantable Medical Devices Directive
NIST GCR 01-817	A Guide to the EU in vitro Diagnostic Medical Devices Directive
NIST GCR 01-823	A Guide to the EU Safety of Toys Directive
NIST GCR 01-824	A Guide to the EU Directive Concerning Liability for Defective Products (Product Liability Directive)
NIST GCR 01-825	A Guide to the EU Directive on General Product Safety
NIST GCR 01-826	A Guide to the EU Directive on Radio Equipment and Telecommunications Terminal Equipment
NIST GCR 01-869	A Guide to the EU Construction Products Directive
NIST GCR 04-870	A Guide to the EU Pressure Equipment Directive

Appendix 6: Lists of International Standards Organizations

The lists in this appendix are based on the information in NIST Special Publication 767, *Directory of International and Regional Organizations Conducting Standards-Related Activities* (edited by Maureen Breitenberg 1989). The International Standards Organizations documented in NIST Special Publication 767 have been segregated into two lists for convenience of reference, one for *treaty* organizations (including United Nations bodies like the International Telecommunications Union [ITU]) and one for *nontreaty* organizations (like the International Organization for Standardization [ISO]). Some additional entries have been gleaned from other references to bring the listings up to date. Common English names have been used in preference to French or other names. Regional Standards Organizations are listed in Appendix 7.

Selected International Standards Organizations (Treaty)

Asbestos International Association (AIA)

Codex Alimentarius Commission (FAO)

Customs Cooperation Council (CCC)

Indo-Pacific Fishery Commission (IPFC)

International Atomic Energy Agency (IAEA)

International Bureau of Weights and Measures (BIPM)

International Civil Aviation Organization (ICAO)

International Commission on Radiation Protection (ICRP)

International Council for the Exploration of the Sea (ICES)

International Frequency Registration Board (IFRB)

International Institute for Cotton (IIC)

International Institute of Refrigeration (IIR)

International Labour Organization (ILO)

International Maritime Organization (IMO)

International North Pacific Fisheries Commission (INPFC)

International Organization for Legal Metrology (OIML)

International Radio Consultative Committee (CCIR)

International Telecommunications Union (ITU)

International Whaling Commission (IWC)

Organization for Economic Cooperation and Development (OECD)

Secretariat of the Convention on International Trade in Endangered Species of Wild Fauna and Flora (CITES)

Universal Postal Union (UPU)

World Health Organization (WHO)

World Intellectual Property Organization (WIPO)

World Meteorological Organization (WMO)

Selected International Standards Organizations (Nontreaty)

African Timber Organization (ATO)

Airport Associations Coordinating Council (AACC)

Association of International Industrial Irradiation (AIII)

Australasian Corrosion Association (ACA)

The Baltic and International Maritime Council (BIMCO)

CAB International (CABI)

Chartered Institution of Building Services Engineers (CIBSE)

International Academy for Quality (IAQ)

International Accounting Standards Committee (IASC)

International Accreditation Forum (IAF)

International Astronomical Union (IAU)

International Commission on Illumination (CIE)

International Commission on Radiation Units and Measurement (ICRU)

International Conference on Large Voltage Electrical Systems (CIGRE)

International Container Bureau (BIC)

International Council of Archives (ICA)

International Council for Building Research, Studies and Documentation (CIB)

International Council on Combustion Engines (CIMAC)

International Electrotechnical Commission (IEC)

International Electrotechnical Commission (IEC) Quality Assessment System for Electronic Components (IECQ)

International Electrotechnical Commission (IEC) System for Conformity Testing to Standards for Safety of Electrical Equipment (IECEE)

International Ergonomics Association (IEA)

International Federation for the Application of Standards (IFAN)

International Federation of Clinical Chemistry (IFCC)

International Federation of Consulting Engineers (FIDIC)

International Federation for Information and Documentation (FID)

International Federation for Information Processing (IFIP)

International Federation of Library Associations and Institutions (IFIA)

IFLA Universal Bibliographic Control and International MARC Program (IFLA UBCIM)

International Gas Union (IGU)

International Institute of Welding (IIW)

International Institution for Production Engineering Research (CIRP)

International ISBN Agency

International Laboratory Accreditation Conference (ILAC)

International Measurement Confederation (IMEKO)

International Microwave Power Institute (IMPI)

International Office of Epizootics (OIE)

International Organization for Standardization (ISO)

International Primary Aluminum Institute (IPAI)

International Radiation Protection Association (IRPA)

International Society of Photogrammetry & Remote Sensing (ISPRS)

International Special Committee on Radio Interference (CISPR)

International Statistical Institute (ISI)

International Union of Biochemistry (IUB)

International Union of Producers and Distributors of Electrical Energy (UNIPEDE)

International Union of Radio Science (URSI)

International Wool Secretariat (IWS)

Appendix 7: List of Regional Standards Organizations

The list in this appendix is based on the information in NIST Special Publication 767, *Directory of International and Regional Organizations Conducting Standards-Related Activities* (edited by Maureen Breitenberg 1989). Some additional entries have been gleaned from other references to bring the listing up to date. Common English names have been used in preference to French and similar names which are included in NIST Special Publication 767. A list of National Standards Bodies appears in Appendix 8.

Selected Regional Standards Organizations

African Regional Organization for Standardization (ARSO)

African Timber Organization (ATO)

Arab Organization for Standardization and Metrology (ASMO)

ASEAN Consultative Committee on Standards and Quality

The Asia-Pacific Economic Cooperation (APEC)

Asia-Pacific Laboratory Accreditation Cooperation (APLAC)

Asia-Pacific Legal Metrology Forum (APLMF)

Asia-Pacific Metrology Program (APMP)

The Asia Pacific Telecommunity (APT)

Association of European Adhesives Manufacturers (FEICA)

Association of European Manufacturers of Fire and Intruder Alarm Systems (EURALARM)

Association of Plastics Manufacturers in Europe (APME)

The Association of Southeast Asian Nations (ASEAN)

Caribbean Common Market Standards Council (CCMSC)

Central American Research Institute for Industry (ICAITI)

The Central American Common Market (CACM)

Central Bureau for Nuclear Measurements (CBNM)

Committee for European Construction Equipment (CECE)

The Common Market of the South (MERCOSUR)

Community Bureau of Reference (BCR)

Conformance Testing Services Programme

Coordinating European Council for the Development of Performance Tests for Lubricants and Engine Fuel (CEC)

The Council for Harmonization of Electrotechnical Standards of the Nations of the Americas (CANENA)

The Disposable Hypodermic and Allied Equipment Manufacturers' Association of Europe (DHAEMAE)

European Aluminum Association (EAA)

European Broadcasting Union Technical Center (EBU)

European Committee for Clinical Laboratory Standards (ECCLS)

European Committee for Electrotechnical Standardization (CENELEC)

European Committee of Manufacturers of Refrigeration Equipment (CECOMAF)

European Committee for Standardization (CEN)

European Computer Manufacturers Association (ECMA)

European Free Trade Association (EFTA)

European Organization for Civil Aviation Electronics (EUROCAE)

European Organization for Quality (EOQC)

European Telecommunications Standards Institute (ETSI)

European Tyre and Rim Technical Organization (ETRTO)

European Workshop on Open Systems (EWOS)

The Free Trade Area of the Americas (FTAA)

The Gulf Cooperation Council (GCC)

Inter-American Association of Sanitary and Environmental Engineering (AIDIS)

The Inter-American Telecommunications Commission (CITEL)

Nordic Committee for Coordination of Electrical Safety Matters (NSS)

The North America Free Trade Agreement (NAFTA)

Organization of American States (OAS)

Pacific Accreditation Cooperation (PAC)

The Pacific Area Standards Congress (PASC)

Pan American Center for Sanitary Engineering and Environmental Sciences (CEPIS)

Pan American Standards Commission (COPANT)

Permanent Secretariat for the General Treaty of Central American Economic Integration (SIECA)

Society for Worldwide Interbank Financial Telecommunications (SWIFT)

Standardization and Metrology Organization for G.C.C. Countries

Standards Promotion and Application Group (SPAG)

Technical Association of the European National Gas Industry (MARCOGAZ)

Appendix 8: List of National Standards Bodies

This list of National Standards Bodies is based on a list published by the National Institute of Standards and Technology at http://www.nist.gov/oiaa/stnd-org.htm, with some updates and revisions. Another list of National Standards Bodies appears on the Standards Engineering Society Web site at http://www.ses-standards.org/displaycommon.cfm?an=1&subarticlenbr=54. That list includes links to the web pages of many of the National Standards Bodies.

Additional information on the addresses of National Standards Bodies appears on the Web site of the World Standards Services Network (www.wssn.net, Structured index–Catalogues link), which provides links to the homepages of the many of the National Standards Bodies.

Australia	Standards Australia (SAA)
Austria	Austrian Standards Institute (ON)
Belgium	The Belgian Institution for Standardization (IBN)
Bolivia	Bolivian Institute of Standardization and Quality (IBNORCA)
Brazil	Associacao Brasileira de Normas Tecnicas (ABNT)
Canada	The Standards Council of Canada (SCC)
China	Standardization Administration of China (SAC)
Colombia	Instituto Colombiano de Normas Tecnicas y Certificacion (ICONTEC)
Czech Republic	Czech Office for Standards, Metrology and Testing
Denmark	Dansk Standard (DS)
Egypt	Egyptian Organization for Standardization and Quality Control (EOS)
Finland	Finland's Standards Association (SFS)
France	Association francaise de normalization (AFNOR)
Germany	Deutsches Institut fur Normung (DIN)
Greece	Hellenic Organization for Standardization (ELOT)
Iceland	Icelandic Council for Standardization (STRI)
India	Bureau of Indian Standards (BIS)
Ireland	National Standards Authority of Ireland (NSAI)
Israel	Standards Institution of Israel (SII)
Italy	Ente Nazionale Italiano di Unificazione (UNI)
Japan	Japanese Industrial Standards Committee (JISC)

Korea (Republic of)	Korean Agency for Technology and Standards (KATS)
Malaysia	Department of Standards Malaysia (DSM)
Mexico	Direccion General de Normas (DGN)
Morocco	Service de Normalisation Industrielle Marocaine (SNIMA)
Netherlands	Nederlands Normalisatie-Instituut (NEN)
New Zealand	Standards New Zealand (SNZ)
Norway	Norges Standardiseringsforbund (NSF)
Peru	Instituto Nacional de Defensa de la Competencia y de la Proteccion de la Propiedad Intelectual (INDECOPI)
Portugal	Instituto Portugues da Qualidade (IPQ)
Russian Federation	Federal Agency on Technical Regulating and Metrology (GOST R)
Saudi Arabia	Saudi Arabian Standards Organization (SASO)
Slovenia	Slovenian Institute for Standardization (SIST)
South Africa	South Africa Bureau of Standards (SABS)
Spain	Asociacion Espanola de Normalizacion y Certificacion (AENOR)
Sri Lanka	Sri Lanka Standards Institution (SLSI)
Sweden	Swedish Standards Institute (SIS)
Switzerland	Swiss Association for Standardization (SNV)
Thailand	The Industrial Standards Institute (TISI)
United Kingdom	British Standards Institution (BSI)
United States	American National Standards Institute (ANSI)
Uruguay	Instituto Uruguayo de Normas Tecnicas
Zimbabwe	Standards Association of Zimbabwe

Index

A